Large-Scale Disasters

Lessons Learned

OECD

ORGANISATION FOR ECONOMIC CO-OPERATION AND DEVELOPMENT

ORGANISATION FOR ECONOMIC CO-OPERATION AND DEVELOPMENT

Pursuant to Article 1 of the Convention signed in Paris on 14th December 1960, and which came into force on 30th September 1961, the Organisation for Economic Co-operation and Development (OECD) shall promote policies designed:

- To achieve the highest sustainable economic growth and employment and a rising standard of living in member countries, while maintaining financial stability, and thus to contribute to the development of the world economy.
- To contribute to sound economic expansion in member as well as non-member countries in the process of economic development. And
- To contribute to the expansion of world trade on a multilateral, non-discriminatory basis in accordance with international obligations.

The original member countries of the OECD are Austria, Belgium, Canada, Denmark, France, Germany, Greece, Iceland, Ireland, Italy, Luxembourg, the Netherlands, Norway, Portugal, Spain, Sweden, Switzerland, Turkey, the United Kingdom and the United States. The following countries became members subsequently through accession at the dates indicated hereafter: Japan (28th April 1964), Finland (28th January 1969), Australia (7th June 1971), New Zealand (29th May 1973), Mexico (18th May 1994), the Czech Republic (21st December 1995), Hungary (7th May 1996), Poland (22nd November 1996), Korea (12th December 1996) and the Slovak Republic (14th December 2000). The Commission of the European Communities takes part in the work of the OECD (Article 13 of the OECD Convention).

Publié en français sous le titre :

Catastrophes de grande ampleur
Leçons du passé

Table of Contents

ISBN 92-64-02018-7
Large-Scale Disasters
Lessons Learned
© OECD 2004

Introduction

by Michael Oborne

This report analyses the economic and social impacts of recent large-scale disasters, and draws some initial lessons for the monitoring and the management of future disasters. The report primarily focuses on restoring trust and securing recovery after a major harmful event has occurred.

The events reviewed are as diverse as the Chernobyl nuclear accident, the Kobe and Marmara earthquakes, Hurricane Andrew, and the 11 September terrorist attacks on New York and Washington. Disasters such as these have in common massive effects on large concentrations of people, activity and wealth. They disrupt multiple vital systems such as energy supplies, transport and communications. Their effects spread beyond the region originally affected and generate widespread anxiety. In some cases, the public expresses distrust of the ability of governments to protect citizens.

The report was prepared between May and July 2003, by a team of specialists from eight OECD Directorates under my supervision. A team of Turkish specialists provided material for Chapter 3.

The text begins with an overview of the important commonalities among these different disaster events (Chapter 1). It is followed by a series of more specialized texts which look carefully at specific sectoral impacts (Chapter 2). The 1999 earthquakes in the Marmara region of Turkey are the subject of a case study (Chapter 3). An annotated bibliography closes the report.

The overview section of Chapter 1 focuses on the following policy messages:

- Governments can – and must – be better prepared to mitigate the economic and social impact of disasters by better planning and coordination across government responsibilities.
- Public trust, as well as consumer and investor confidence are key elements to ensure rapid and systemic recovery; these elements need to be strengthened through credible communication and effective action of both the public and private sectors.
- Governments need to work more closely in partnership with the private sector, which has key roles to play in disaster prevention, preparedness, response and recovery.
- Major disasters and harmful events can have multiple international dimensions, and these call for more systematic international co-operation.

These policy lessons are drawn from extensive OECD work on disaster-related issues.

The sectoral notes in Chapter 2 provide insights into recent OECD work.on risk and disaster impact by analyzing eight specific issues:

- Measuring the impact of large-scale disasters.

- Economic recovery from past disasters.

- Impacts on public finances.

- The consequences for financial and insurance markets.

- Disaster management through insurance.

- Compensation issues.

- Housing and community reconstruction.

- Lessons learned from nuclear accidents.

Chapter 3 of the report deals with the Turkish response to the Marmara earthquake disaster in some detail. An annotated bibliography of recent OECD publications related to disaster management is included to guide the reader toward recent economic literature in this field.

This material is now made available to a wider public. The issues raised in this report concern the welfare of citizens, and at the same time provide further reflection on the ways in which individuals, acting both through the public sector and the private sector, can influence and shape new ideas on risk management. Risk is a multi-sectoral, multidisciplinary set of issues that must increasingly be addressed across traditional administrative divides and government responsibilities.

ISBN 92-64-02018-7
Large-Scale Disasters
Lessons Learned
© OECD 2004

Chapter 1

Lessons learned

by Reza Lahidji

Many OECD countries have been affected by major harmful events in recent years. The considerable human and economic costs of such events and the repercussions they might have for the global economy have become recurring causes for concern.

Given its intergovernmental and multidisciplinary nature, and its experience in risk and disaster management in a variety of fields,[1] the OECD is well positioned to analyse the impact of major disasters on societies and economies, and to identify good and bad practices in the response and recovery phases. Building on the past work of the Organisation, and on the in-depth studies included in the following chapters of the book, this chapter proposes a set of policy lessons learned from the impact and management of major harmful events in various parts of the world.

The analysis focuses primarily on large-scale disasters such as the Chernobyl nuclear accident, the Kobe and Marmara earthquakes, Hurricane Andrew and the 11 September terrorist attacks. Detailed data on these five disasters is presented in boxes. It is important to emphasise here that available data on megadisasters of this kind tend to be quite scarce and therefore provide an incomplete picture of the event in question. In particular, relatively little work in conducted on measuring the economic consequences of major disasters. Quite often, there are also problems of consistency between the various sources of information.

These various events differ in many respects, for example in the extent of damage, the context in which they occurred and the responses of governments. But they also have similarities, in particular in the way they spread and affected large parts of our societies and economies.

Useful lessons from past disasters can therefore be drawn, although they should not be seen as foolproof recipes for handling future events. In all likelihood future disasters will differ sharply from previous ones and, even if well prepared, governments will still have to face unexpected developments. But lessons from the past can provide a framework for facilitating and improving disaster response and recovery.

Lesson 1. Governments can and must be better prepared to mitigate the economic and social impact of disasters.

In many cases, the scale and particular features of past disasters have taken governments by surprise. For instance, none of the countries affected by radiological contamination after the Chernobyl accident had made provisions or set up operational plans for an event of such proportions.

Disasters can have substantial economic impacts

In addition to their terrible human toll, large-scale disasters can cause considerable economic damage, in the order of magnitude of one percentage point of total wealth or several percentage points of GDP. Today, disasters generally affect or threaten large conurbations with high concentrations of population, economic activity and wealth.[2, 3] Some of these areas are regional or even global centres for crucial sectors (*e.g.* finance in Tokyo or New York, international financial governance in Washington, transport hubs and integrated communications centres). Critical infrastructures may be damaged, and systems upon which society and the economy depend (health care, telecommunications, transport, energy supply, etc.) severely disrupted.

The negative economic consequences can be substantial, especially if there is a threat of repetition of the disaster (terrorism, earthquakes, radiological contamination). Disasters can have a short-term destabilising effect on the economy because of their impact on consumer and business confidence, the liquidity needs they may create in the financial sector, and other sectoral imbalances they may engender.

Decision making must be flexible and responsive

Because accidents are unpredictable, preparedness cannot simply consist of guidelines and procedures to be followed. Decision making must be capable of reacting to the unexpected in a timely and effective manner.

In the aftermath of a disaster, governments face considerable pressure to intervene: to reduce or contain persisting dangers, compensate victims, clean up and reconstruct damaged areas, provide temporary shelters, subsidise affected industries and local governments, prevent liquidity crises, and restore confidence. In some cases in the past, the fiscal costs of disaster response have exceeded 1% of GDP for several years (see Key issue 2 in Chapter 2).

Box 1.1. **The Hyogo-ken Nanbu earthquake**

Initiating event:	Earthquake of magnitude 7.2.
Date:	17 January 1995
Location:	Hyogo prefecture, Japan. *Kobe, one of the country's major ports (30% of total freight) was stricken.*
Fatalities:	6 430.
Impact on housing:	105 000 houses destroyed, 144 000 damaged, 310 000 persons evacuated in shelters.
Economic damage:	USD130 billion. *80% decrease in freight transport (imports + exports) in the port of Kobe in the weeks following the disaster. Water, gas, and electricity supply, railway and road transport, telecommunications and manufacturing production were severely disrupted. For instance, water supply to 1 230 000 houses and to industrial water supply to 289 plants was interrupted. Electricity supply to 2 600 000 customers was stopped.*
Insured loss:	USD 3 billion (indexed 2002) excluding life and liability insurance losses.
Fiscal impact:	USD 70 billion in 1994, USD 100 billion in 1995 (1.3% and 1.9% of GDP respectively).
Source:	Various sources cited by the OFDA/CRED database (*www.cred.be/emdat*); Japan National Land Agency, *Disaster Prevention White Paper*, 1998; Swiss Re, *Sigma*, No. 2, 2003.

LARGE-SCALE DISASTERS – ISBN 92-64-02018-7 – © OECD 2004

Government intervention can have negative effects
if margins of manœuvre are inadequate
or interventions not carefully tailored

Such interventions can generate two types of adverse effects:

- at the macroeconomic level, they may greatly increase public indebtedness and fuel inflation;
- at the microeconomic level, they may have distortionary effects and generate disincentives.

It is therefore important that governments preserve substantial margins for action in the budgetary area, and possibly also in the monetary area. In addition, as discussed in the following lessons, governments need to design their interventions carefully in terms of scope and duration in order to address their possible distortionary effects.

Incomplete information calls for special decision-
making capacities and tools

During and immediately after an event, information comes piecemeal and is often hard to put together and interpret. Vital elements of the picture are scattered across central and local government, law enforcement and regulatory authorities, corporations and other stakeholders in the sectors concerned, from citizens to hospitals.

Decisions therefore usually have to be made on the basis of incomplete information, in a context of utmost urgency, and with considerable human, economic and political stakes. Protection and evacuation decisions are a good illustration. Experience shows that to be effective in such conditions, disaster management should rely not on detailed procedures, but rather on a responsive decision-making structure. All organisations and individuals involved in emergency response need to have clearly identified roles and responsibilities, and should be trained to communicate and to co-ordinate their actions. In addition, decision analysis models using probabilistic methods can be a precious tool for decision makers in emergency situations, for instance by defining levels at which intervention should be triggered.

Box 1.2. **The Marmara earthquakes**

Initiating events:	Two earthquakes of magnitude 7.6 and 7.2 respectively.
Date:	17 August and 12 November 1999.
Location:	Kocaeli, Sakarya and Duzce districts of northwestern Turkey.
	The affected area concentrates 23% of the population, 34% of Gross National Product, and 46% of total industrial output. The Kocaeli-Sakarya region, in particular, is considered as the industrial heartland of the country, and provides 16% of total government revenues.
Fatalities:	18 373.
Impact on housing:	109 000 homes/business collapsed, 249 000 homes/ business damaged, 600 000 homeless.
Economic damage:	Between USD 9 and 13 billion according to the State Planing Organisation, between USD 6 and 10 billion according to the World Bank.
	GNP decreased by 7.6% year-on-year in the third quarter and by 4.9% in the fourth quarter of 1999. 140 000 people were left jobless in the affected area.
Insured loss:	USD 1-2 billion (indexed to 2002).
Fiscal impact:	USD 1.8 billion in 1999, USD 4.2 billion in 2000 (1.0% and 1.9% of GNP respectively).
Source:	OECD, "Economic effects of the 1999 Turkish earthquakes: An interim report", *Economics Department Working Papers*, No. 247, 2000; Akgiray, V., G. Barbarosoglu and M. Erdik, "The Marmara Earthquakes in Turkey", Chapter 3 of this book; Swiss Re, *Sigma*, No. 2, 2003.

LARGE-SCALE DISASTERS – ISBN 92-64-02018-7 – © OECD 2004

Lesson 2. The public's trust and consumer and investor confidence are key ingredients of recovery; they need to be strengthened through credible communication and effective action.

In societies that have reached very high safety and security standards, large-scale disasters can cause considerable anxiety and loss of confidence among the population.

Credible communication is key

In the aftermath of a harmful event, there is a strong demand for information. Independent media collect and interpret data and present them to the public, often in real time. Governments have little control over the way information is handled by the media and received by the public. Their ability to communicate freely may be hampered by the need to withhold critical information, for instance to protect potential targets and prevent malevolent actions. Disaster situations provide fertile ground for disinformation and rumours, possibly leading to panic and stigmatisation of certain regions or social groups. On the other hand, people respond to disasters in a rational and responsible manner when official sources are credible and trusted.

The public's confidence in risk management authorities is therefore key to disaster control and recovery.

Public trust requires long-term dialogue with stakeholders

Building public trust requires transparency and openness in communication over time, an element that has often been neglected in the past. In the reconstruction phase, for instance, governments have tended to do too much too quickly, favouring large recovery expenditures over the longer-term needs and safety of the population. Stakeholders have seldom been involved in the assessment of damage or needs (see Key issue 6 in Chapter 2) – *i.e.* in answering questions such as:

● How have living conditions and social relations been affected?

● How likely are they to recover thanks to local and private initiatives?

● How can such initiatives be backed by public policies – in the re-building of infrastructure, for example?

● Are there opportunities to improve the facilities and infrastructure being replaced?

Lack of information to exposed populations and their non-involvement are part of the reason why, fifteen years after the Chernobyl catastrophe, public distrust of government, lack of confidence in the future, and psychosomatic illnesses still impose considerable costs on the Ukraine and Belarus governments.

If disasters reveal flaws in risk handling under-
and over-reaction must be avoided the real cause
of failure investigated and adequate action taken

Often, large-scale disasters reveal important flaws in the way a risk has been handled: *e.g.* a significant threat has been underestimated; early warnings have not worked; safety norms and procedures have not been followed. In terms of the regulatory reponse to disasters, one of two polar (but equally inappropriate) attitudes often emerges: to continue as if nothing had happened, or to search for someone to blame and tighten command-and-control procedures.

A disaster heightens awareness among all stakeholders. The appropriate course of action, then, is to involve them in designing better risk management through co-operation, better use of technological tools, improved economic incentives, etc. This entails trying to answer questions such as:

● Are there any precursors to the occurrence of a hazard, and can monitoring detect them?

● Did the occurrence of a hazard correspond to earlier assessments?

● How did the disaster spread, and whom did it affect?

● How did people react, and were warning signals received?

● Were there any unexpected factors of vulnerability?

● Which social and economic trends contributed to creating vulnerabilities, and can they be better managed?

● Did any protective measures fail, and if so, why?

● Were there effective incentives to avoid or mitigate risk?

LARGE-SCALE DISASTERS – ISBN 92-64-02018-7 – © OECD 2004

Box 1.3. **The Chernobyl nuclear accident**

Initiating event:	Meltdown of Unit 4 of the Chernobyl nuclear power plant.
Date:	26 April 1986.
Location:	Chernobyl, Ukraine.
Fatalities:	31 immediate victims.
	23 million people lived in the contaminated area inside Ukraine. 135 000 people were reported affected.
Economic damage:	USD 2.8 billion.
Insured loss:	–
Fiscal impact:	USD 18 billion during the Soviet Union era (1986-1991). The cumulative expenditure related to Chernobyl in Belarus from 1991 to 2001 represent 20% of the 2001 GDP. Both the Ukraine and Belarus still spend close to 2% of their GDP in Chernobyl related programs.
Source:	CRED/OFDA database; World Bank, Belarus – Chernobyl Review, 2002.

Investigating the sources of a disaster and then engaging adequate corrective actions increase public trust in the government's capacity to handle future risks.

Lesson 3. Governments need to work in partnership with the private sector, which has key roles to play in disaster prevention, preparedness, response and recovery.

The private sector has a crucial role to play in rebuilding the areas affected by a disaster and restoring economic dynamism.

The private sector, unless devastated, has the capacity to restore economic performance

---

Past recovery patterns show that cities and regions that have suffered a disaster often bounce back after a few months or years, except in cases of extreme devastation (*e.g.* involving radioactive or chemical contamination). The private sector has the capacity to invigorate local economies, providing it finds adequate incentives.

Public measures aimed at supporting affected industries and regions or increasing security can distort competition and hamper trade. Therefore, they need to be carefully designed, strictly regulated, and limited in scope and duration.

Systematic assistance can perpetuate "victim status"

Public interventions aimed at helping victim and providing compensation can also entail moral hazard problems (see Key issue 1 in Chapter 2). Humanitarian assistance is imperative in the wake of a large-scale disaster, and governments often provide more substantial compensation as an expression of national solidarity. But if compensation becomes systematic – even in cases where people have not taken basic measures to protect themselves – it risks paving the way for future disasters. In the same vein, provision of permanent social assistance to disaster victims can create the incentive to ignore or downplay hazards (*e.g.* in a contaminated area) and perpetuate "victim

LARGE-SCALE DISASTERS – ISBN 92-64-02018-7 – © OECD 2004

status". Provision of public compensation or social assistance must also encourage proactive behaviour, integrate strong incentives for prevention (to follow building codes, relocate, etc.), link benefits to actual harm, and be gradually phased out.

Responsibilities of actors in risk management should be better defined

The respective roles and responsibilities of the public and private sectors in risk management have undergone significant transformation in OECD countries in recent decades. In the wake of changes in society, regulatory reform and privatisation, centralised command-and-control procedures regulating safety have lost their effectiveness. A number of recent accidents and disasters have revealed gaps in the way risks are handled. Countries need to clarify the respective responsibilities of the actors and to adapt legal frameworks – liability laws in particular – to this rapidly changing environment. Some countries have started doing this in privatised public utilities and in certain segments of the transport sector.

Soft regulation can do more to improve resilience and reduce vulnerability

Public and private partnerships can also help reduce vulnerability and increase resilience by improving "soft regulations". For instance, in the aftermath of 11 September, it appeared that the business continuity of certain banks had been hampered by two factors: the close proximity of backup facilities to the primary sites, and insufficient attention to updated and operational backup procedures. The banking industry has started to tackle these issues.[4] A move towards self-regulation has been encouraged and supported by the competent regulatory bodies (the Federal Reserve Board, the US Securities and Exchange Commission and the Office of the Comptroller of the Currency) through new supervisory guidance and examination procedures.[5]

Finally, public/private partnerships in the financial compensation process have to be strengthened (see Key issue 5 in Chapter 2).

Mega-risk burden sharing requires new public/private co-operation

The recurrence of very large-scale disasters in recent years has shown the limits of traditional risk-sharing mechanisms. Insurance against disasters has become more comprehensive in the past decades, and nowadays covers a substantial share of the damage incurred in most OECD countries. However, insurance cover varies greatly among OECD countries and for the different types of catastrophic risks. This carries serious economic and budgetary consequences for the management and compensation of disasters. Moreover, substantial losses experienced after events such as Hurricane Andrew, 11 September and the 1994 Northridge (California) earthquake have called into question the industry's capacity to deal with such mega-risks.[6] (Re)insurers have sometimes refused cover for certain major risks, at least at affordable prices. Cover against terrorist acts was temporarily suspended in the United States after the World Trade Center attacks. Available evidence in insurance markets shows that the global insurance and reinsurance industry may not be able to withstand another shock of the magnitude of 11 September (see Key issue 3 in Chapter 2).

Specific risk-sharing mechanisms involving insurance and reinsurance companies, pooling structures, capital markets, governments and even international structures may need to be designed in order to provide cover against catastrophic terrorism and possibly other mega-risks.[7]

Box 1.4. **The 11 September terrorist attacks**

Initiating event:	Airplanes hijacked and crashed by terrorists.
Date:	11 September 2001.
Location:	New York, Washington DC and Pennsylvania, United States.
Fatalities:	Over 3 000.
Economic damage:	USD 120 billion.
	Destruction of physical assets was estimated at about USD 16 billion in the national accounts. Close to 200 000 jobs were lost or relocated. Business and consumer confidence indexes fell sharply. The unemployment rate rose from 5% in September to 5.8% in December. However, the GDP fell only moderately in the third quarter, and rose again in the fourth.
Insured loss:	USD 19 billion excluding life and liability insurance losses, USD 40-50 billion total.
Fiscal impact:	–
Source:	OECD, "The Economic Consequences of Terrorism", *Economics Department Working Papers*, No. 334, 2002; Swiss Re, *Sigma*, No. 2, 2003.

Lesson 4. Major disasters or harmful events can have global implications and call for international co-operation.

Disasters can overwhelm the response capacities of any single country

The scale of a single or multiple harmful event can easily overwhelm the capacities of a single country to prevent the global spread of a threat. Indeed, various hazards can be global in nature (a disease, an environmental impact, etc.). In many cases, nation-based risk management strategies need to be complemented by international co-operation. Such mechanisms exist for specific areas, some sectoral (*e.g.* nuclear energy), others broadly economic (*e.g.* the IMF's role in stabilisation). Links need to be strengthened between the domestic and global levels of disaster management and recovery.

Increased international co-operation may include:

- Information and knowledge sharing.
- Co-ordination of national initiatives.
- Design of international tools for disaster management.
- Binding agreements.

Information sharing is crucial...

There are cases (*e.g.* involving biological, chemical or radiological contamination) in which any situation with public health implications – even if it initially affects only a restricted geographic area – will be of interest to the governments of other countries (see Key issue 7 in Chapter 2).

... but still mostly depends on the surveillance capacity of individual countries

However, global tracking and transmission of information are still in their infancy for most major hazards. Even where comprehensive international surveillance and monitoring structures have been developed, these consist of networked national and regional systems. As recently demonstrated by the SARS epidemic, effective global tracking of hazards greatly depends on the surveillance capacity of individual participating countries, and on their willingness to co-operate. Technology and knowledge transfers to and capacity building in developing countries are therefore a necessary – though not sufficient – element of any global disaster containment strategy.

Unco-ordinated reactions to disasters have at times contributed to disruptions in trade, investment and travel. Co-ordination must be a policy goal with institutional underpinnings, such as *ex ante* international agreements. These agreements could be modelled along the lines of, say, the radiological criteria that apply to trade in food.

International risk-sharing mechanisms may be needed for "mega-risks" (see Key issue 4 in Chapter 2). Such mechanisms already exist in the areas of maritime oil and chemical pollution and nuclear liability.

International risk- and burden sharing tools should be developed

A number of international co-operative platforms providing financial support to affected countries could also be further strengthened. This is notably the case with relief efforts directed to regions that have suffered by natural catastrophes, where poor co-ordination of information and logistics frequently leads to over- or under-response to the disaster.

Finally, the globalisation of disasters raises a number of questions that are beyond the scope of this study. What triggering mechanisms could be designed to enable disaster management to shift from the national to the international level? What can be done to ensure that countries co-operate? On what basis would countries contribute to a compensation scheme? Should such a scheme cover a single type of disaster, a category of comparable (*e.g.* natural) disasters, or all major risks?

Answers to these questions will need to be explored in appropriate multilateral fora.

Box 1.5. **Hurricane Andrew**

Initiating event: Hurricane.

Date: 24 August 1992.

Location: Southern Florida peninsula and south-central Louisiana, United States, and northwestern Bahamas.

Fatalities: 38.

Economic damage: USD 25 billion.

In Florida, more than 135 000 homes were destroyed or damaged, 160 000 people were left homeless and 86 000 lost their jobs. In Louisiana, an additional 21 000 homes were destroyed or damaged.

Insured loss: USD 20 billion excluding life and liability insurance losses.

Fiscal impact: –

Source: The United States National Oceanic and Atmospheric Administration's press release for the 10th anniversary of Hurricane Andrew, 22 August 2002; Swiss Re, *Sigma*, No. 2, 2003.

Notes

1. For a general survey of these activities, see the annotated OECD Bibliography.

2. The world urban population exceeds 3 billion today, compared to 1 billion in 1960. By 2030, it is expected to reach 5 billion. See United Nations, *World Urbanisation Prospects 1999*, Department of Economic and Social Affairs, Population Division, 2001.

3. Forty of the fifty fastest-growing urban centres in the world are located in earthquake-endangered areas. See Randall, J.B., D.L. Turncotte and W. Klein (eds), *Reduction and Predictability of Natural Disasters*, Santa Fe Institute Studies in the Science of Complexity, Addison-Wesley, 1996.

4. Ferguson, R.W., "A Supervisory Perspective on Disaster Recovery and Business Continuity", Remarks before the Institute of International Bankers, 4 March 2002, Federal Reserve Board, Washington DC.

5. Board of Governors of the Federal Reserve System, "Interagency Paper on Sound Practices to Strengthen the Resilience of the US Financial System", 2003.

6. These three events have each caused close to USD 20 billion of insured losses, in property and business interruption alone. See Swiss Reinsurance Company, *Natural Catastrophes and Man-made Disasters in 2002*, 2002.

7. Nuclear damage, marine oil pollution and, in some countries, terrorist acts are already subject to specific multi-pillar risk-sharing arrangements.

ISBN 92-64-02018-7
Large-Scale Disasters
Lessons Learned
© OECD 2004

Chapter 2

Key issues

Key issues

1. Economic recovery from past disasters

by Patrick Lenain

Large-scale disasters affect national economies through various channels. They severely disrupt business activity, undermine investor and consumer confidence and cause chains of other disruptions. These impacts are not necessarily very large or permanent on a national scale. But they can be very significant at the local level and notably disrupt local economic activity. For instance, regions and cities stricken by earthquakes typically suffer from a sharp drop in output and private consumption. Zones contaminated by nuclear radiations may succumb to unemployment and poverty for a protracted time period. Governments are therefore often under pressure following large-scale disasters to intervene and facilitate the economic recovery. Rebuilding public infrastructure and providing short-term humanitarian assistance are appropriate elements of response. Beyond this, government interventions need to be handled carefully, and the case for permanent government intervention is unclear. For instance, generous programmes of social assistance, if permanent, can reduce economic incentives. Large payments to victims and evacuees may create disincentives and result in "victim and dependency mentality" that might hamper the economic rebound.

How do large-scale disasters impact the economy?

Local impact

Large-scale disasters affect local and national economies through a variety of channels. *At the local level, the effects can be very large.* For instance, sales of department store in the city of Kobe fell by 74 per cent in February 1995 compared to the previous year because many supermarkets were destroyed during the earthquake. Many production facilities were lost, resulting in a sharp drop of production and employment. Following the Chernobyl accident, activity in contaminated areas fell sharply because the population was resettled to areas with lower levels of radiations; food production and forest-exploitation were cut back sharply.

Although the local economic impact of disasters is large, the rebound is often rapid in OECD countries.[*] Available evidence and economic theory suggest that

---

[*] Available evidence suggests that the effects of large-scale disasters are much more pervasive in developing countries because these economies lack the ressources necessary to reconstruct. In these countries, low-income people are particularly at risk (Freeman *et al.*, 2003).

local economies recover rather quickly from the consequences of disasters (Kepenek, Yetkiner and Zon, 2001). An economy which keeps the bulk of its human capital and productivity level after a large-scale disaster will find that damaged areas will be rebuilt after a few years. Local economic rebound requires the reconstruction of housing, buildings, public infrastructure and production facilities, which typically takes a few years and is financed by the settlement of insurance claims. In Kobe, eight years after the earthquake, the local economy appeared to have achieved nearly full recovery, even though department store sales were still about 10 per cent below the level achieved before the earthquake. Following the Chernobyl accident, economic activity moved to a new settlement, which shows signs of healthy rebound (see Box 2.1 on Slavutych).

Box 2.1. **Local economic recovery in the city of Slavutych**

The city of Slavutych was built after 1986 to resettle people living near Chernobyl. Despite the closure of the nuclear plant, the city enjoys relatively prosperous economic conditions. The unemployment rate is lower than the national average and business creation is growing. This is partly due to the status of the city as a special economic zone, and notably the tax exemption granted for several years to new businesses. The special attention paid by the government and the international community seems to bear fruit.

In areas where large quantities of nuclear material fell (the so-called "exclusion zone" covering 3 000 km^2), by contrast, there is no activity because the populations have been more or less permanently evacuated and no agricultural activity is permitted. In addition, products coming from lesser-contaminated areas (approximately 100 000 km^2), especially food products, are likely to be contaminated themselves and be unsafe for consumption.

Similarly, regions where terrorist activities persist (*e.g.* recurrent car-bombing or hostage-taking) can be undermined durably. Research on the economy of the Basque region suggests for instance that GDP is about 10 per cent lower due to the various forms of terrorist activities and their impact on investment (Abadie and Gardeazabal, 2001).

National impact

Large-scale disasters have limited effects at the aggregate level. Because they destroy physical capital, disasters *reduce the national stock of wealth*. However this impact is fairly small on a national scale. The amount of damage that occurred during the Kobe earthquake is estimated at USD 130 billion, which is equivalent to only 10 per cent of the country's annual capital formation. Losses resulting from the Marmara earthquakes of 1999 are estimated to have ranged between USD 3 and USD 6 billion according to the World Bank, the equivalent of 6 to 12 per cent of the country's annual investment. Such estimates are of course subject to large uncertainties, notably related to the damage and losses occurred by the enterprise sector.

Disasters also reduce the number of jobs and lower economic activity in the short-term. This reflects damages to business premises in the affected region, loss of lives and migration. The impact on GDP of past disasters is difficult to evaluate, but is unlikely to have been large. According to OECD (2001), the Marmara earthquake is estimated to have reduced the growth of Turkey's GDP by 1 percentage point reflecting disruptions to supply (loss of physical capital and labour force) and demand (loss of inventories, temporarily depressed consumption and investment activity, interrupted linkages between enterprises, disruption to domestic and international transportation). Reconstruction, resettlement and recovery activities *can actually boost activity* and increase GDP growth within a relatively short period. The terrorist attacks of 11 September in New York were for instance followed by an increase in GDP during the fourth quarter of 2001, interrupting the recession.

Sectors of activity are affected differently by large-scale disasters. Construction companies obviously benefit from earthquakes, because they are hired for programmes of reconstruction on a large magnitude. In contrast, the *aviation* sector was deeply undermined by the terrorist attacks of 11 September through various channels. The *insurance sector* typically suffers heavy losses, and *commercial banks* may also see a sharp increase in non-performing loans.

Finally, disasters can have a number of other direct and indirect effects. Large-scale reconstruction can for instance increase *inflation*, notably if the demand for new housing exceeds supply. In the case of Turkey, the earthquake was followed by an increase in the *trade deficit*, because the country had to import products in short supply and exports were negatively affected. The Marmara earthquake also temporarily increased Turkey's *risk premium* on international financial markets, which contributed to *higher debt service payments* and raised the current account deficit. In the case of the 11 September attacks, the *stock market* was closed for nearly one week, and indices fell sharply after the reopening of markets. Both business and consumer *confidence* fell sharply. These problems were however temporary, and both the stock markets and consumer rebounded after a few months.

What is the case for government intervention?

Local economies

Apart from the reconstruction of public infrastructure, governments have used two types of interventions at the local level after large-scale disasters.

First, they have provided assistance targeted to people who lost their housing. After the Marmara earthquake, for instance, the government organised the establishment of tent villages to shelter those who lost everything and could not find another accommodation. As cold winter came, the government financed the construction of prefabricated settlements. These programmes of shelter were temporary and were phased out as housing reconstruction was completed. *Second*, governments have provided temporary financial help to small businesses. For instance the Japanese government provided various financial facilities to help the SME sector, such as "Restoration Special Loans" as well as tax reductions and tax deferrals. The Ukrainian government granted the status of "special economic zone" to the city of Slavutych.

Temporary financial support has been deemed warranted to help people cope with the consequences of a disaster and prevent extreme cases of poverty and distress. *Permanent social assistance may however have adverse effects.* For instance the Government of Belarus provided people living in the most contaminated areas compensation as well as various benefits such as cash payments, free medical treatment, free transportation, discounts on utility and rental payments, free access to sanatorium and recreation facilities and education privileges. According to the World Bank (2002), these various forms of social assistance have encouraged the emergence of a "victim mentality", which reduces the incentive for affected people to seek a new job, resettle to a new area and be proactive in starting a new life. Large groups of people thus live in poverty in contaminated areas, but do not wish to move out of these areas, at least partly because they would lose their social benefits. This negative consequence of government support may be of particular concern when social assistance is linked to the place where victims live (*i.e.* disaster areas) rather than their health status (*i.e.* medical problems or the need for medical surveillance).

National economy

At the national level, governments have traditionally focussed on comforting populations struck by the magnitude of disasters. They have endeavoured to reassure consumer and businesses that they were safe and that their properties were looked after (for instance regarding the safety of deposits and investments held by the banking system). Macroeconomic policy tools have been used to bolster confidence. After 11 September, large scale injections of liquidity in the banking have helped preserve the integrity of the financial system and reassure depositors, despite the physical destruction of building blocks in New York's financial district. Restoring confidence at the nationwide level, together with appropriate macroeconomic actions, may thus help avoid a negative national economic impact.

Conclusion

Because each disaster is different, it is difficult to draw universal lessons on economic recovery following large-scale disasters. The following conclusions seem nonetheless to apply to many cases. *First*, the economic impact of large-scale disasters is significant at the local level, with physical destruction on large magnitude, losses of lives, disruption or interruption of business output and sharp declines in consumption. The available evidence suggests however that local economies tend to recover rather quickly from disasters. The city of Kobe, the Turkish Marmara region and the Ukrainian city of Slavutych had recovered a few years after being struck by disasters. *Second*, permanent social assistance, especially if linked to the area of residence, can reduce incentives among victims and evacuees to start a new life. Social assistance targeted to individuals with medical concerns is preferable to less specifically targeted benefits, such as by geographic region. *Third*, at the national level, large-scale disasters have had limited effects. The most important task of the authorities is to restore confidence, so that consumers and investors can resume their normal routines.

References

Abadie and Gardeazabal, 2001, *The Economic Costs of Conflict: A Case-Control Study for the Basque Country*, NBER Working Paper No. w8478.

Freeman, P. K., M. Keen and M. Mani (2003), *Dealing with Increased Risk of National Disasters: Challenges and Options*, IMF Working Paper WP/03/197, Washington D.C.

Kepenek Y., H. Yetkiner and A. Zon (2001), *Eartquakes and Growth*, mimeo.

OECD (2001), *Economic Survey of Turkey*, Paris (Chapter IV. Policy implications of the 1999 eartquakes).

World Bank (2002), *Belarus – Chernobyl Review*, Report No. 23883-BY, Washington DC.

Key issues

2. Budgetary impact

by Patrick Lenain

After large-scale disasters, governments are generally expected to help victims and businesses deal with their trauma and losses. Governments are expected to finance part of the reconstruction, clean-up and recovery effort and look after the health of victims. This note examines the composition of budgetary spending related to disasters, the order of magnitude and finally draws some lessons.

Past experience

Large-scale disasters can be costly, including for governments which are expected to help local economies and victims recover. The *direct* budgetary assistance can take different forms:

- Containment of the disaster to prevent further risk (*e.g.* construction of a temporary sarcophagus around the Chernobyl nuclear plant).
- Clean-up and immediate recovery operations (*e.g.* cleanup of lower Manhattan after 11 September).
- Construction of shelters for homeless victims (tent communities and villages of prefabricated houses in Turkey).
- Reconstruction of damaged public transportation facilities, roads, railways, ports and airports (*e.g.* infrastructure destroyed by the Kobe earthquake).
- Financial support for the reconstruction of eligible housing (*e.g.* 55 to 75 per cent of affected housing in Turkey).
- Compensation payments and cash benefits to victims (*e.g.* payments to families who lost a parent during the 11 September attacks; assistance to people living in radioactive areas in Belarus).
- Financial support to help small business recovery (*e.g.* Kobe earthquake) or to compensate a specific sector particularly affected by the disaster (*e.g.* compensation payment to the airline industry after 11 September).
- Budgetary transfers to local governments (such as City of New York or City of Kobe) to help them mitigate the impact on municipal and state budgets.

In addition, public finances can be affected via *indirect* ways:

- Losses of tax revenues normally collected in disaster areas.
- Increased interest payments due to higher risk premium on government debt (Turkey).

LARGE-SCALE DISASTERS – ISBN 92-64-02018-7 – © OECD 2004

● Measures to combat the factors that caused the disaster (such as anti-terrorism measures after 11 September).

The fiscal cost of disasters

There is no internationally-agreed methodology to measure the fiscal impact of large-scale disasters, which makes cross-country comparison difficult. In addition, some governments tend to exaggerate the budgetary impact of disasters for various reasons. It is therefore beyond the ambition of the following paragraphs to propose a comparison of the fiscal costs associated with various disasters. Instead, the fiscal impacts of three disasters are reviewed individually.

Marmara earthquake. The fiscal impact of the 1999 Turkish earthquakes has been well documented by various international organisations (see for instance OECD, 2001). The earthquake was expected to increase Turkey's budget deficit by USD 1.7 billion in 1999 (about 1 per cent of GNP) and USD 4.2 billion in 2000 (about 2 per cent of GNP). The costs borne by the budget included the following: extra consumption and transfer spending for the relief effort and additional social security spending linked to death and disability benefits; credit subsidies and tax deferrals/losses for affected businesses and individuals; new investment spending for the construction of temporary housing and gradual reconstruction of housing and infrastructure; repairs to damaged transport and communication infrastructure, schools and hospitals. A large part of this fiscal cost was met with official foreign financing managed through the World Bank-financed project implementation unit. The remainder was financed with one-off tax measures announced in the context of an "earthquake package".

Kobe earthquake. The Japanese government estimates that the fiscal cost linked to the Kobe earthquake was USD 70 billion in 1994 (over 1 per cent of GDP) and USD 100 billion in 1995 (about 2 per cent of GDP). This included a variety of government programmes such as repairs to roads, water and sewer systems, harbour facilities and schools. Earthquake victims were made eligible for relief grants, low-cost loans and tax relief. To cover the cost of reconstruction, the government issued bonds in anticipation of tax revenue drops. Taking into account earthquake-related reductions in corporate and personal income taxes, the Finance Ministry anticipated a drop of ¥500-600 billion (USD 6-7 billion) in tax revenues for the fiscal year.

Chernobyl accident. The fiscal cost of the Chernobyl accident is surrounded by controversial debates and an accurate estimate will probably never become available. During the period 1986-1991, the accident was dealt with by the Soviet Union, which according to some estimates spent the equivalent of USD 18 billion on rehabilitation, of which 35 per cent on social assistance to affected people and 17 per cent on resettlement (UNDP, 2002).

After the dissolution of the Soviet Union, the fiscal cost of dealing with the aftermath of the accident became the responsibility of the governments of Belarus, Russia and Ukraine. Belarus has by far the largest contaminated areas and therefore has to deal with large population resettlements and health problems. Ukraine has to deal with the remains of the Chernobyl nuclear plant and the construction of a new sarcophagus (this question is not addressed in the present annex). Russia has relatively small contaminated areas but has a population of victims exposed to radiations, including so-called "liquidators" who dealt with the containment and cleanup efforts after the accident. Excluding the construction of a new shelter around the plant, the largest fiscal cost is today borne by Belarus. According to the World Bank (2002), the present value of resources spent from the Republican budget of Belarus since 1991 amounts to about 20 per cent of the 2001 GDP. Belarus now devotes about 1-2 per cent of GDP to Chernobyl-related programmes. The largest fiscal cost came initially from capital expenditure, as the government was financing the cost of constructing new settlements and their public infrastructure (schools, hospitals, roads, utilities, etc.). The bulk of the cost now stems from the payment of social benefits and compensation to people living in contaminated areas, invalids and other victims. In addition, the government finances a number of targeted projects, for instance to evaluate health problems and to monitor radiation levels. In Belarus and Ukraine, the recurrent cost of Chernobyl-related spending has put additional pressures on already strained budgets. Foreign assistance from international institutions and non-governmental organisations has therefore played a useful role.

Conclusion

Overall, the fiscal impact of large-scale disasters has in the past been in the range of 1-2 per cent of GDP for a few years (following earthquakes) or many years (following the Chernobyl accident). Programmes to deal with disasters can be difficult to finance for countries that already had a stretched budget. It is therefore important to keep some room for manœuvring in the budget so as to respond quickly and in a non-inflationary manner to disasters.

References

OECD (2001), Economic Survey of Turkey, Paris.

UNDP (2002), The Human Consequences of the Chernobyl Nuclear Accident – A Strategy for Recovery, mimeo.

World Bank (2002), Belarus – Chernobyl Review, Report No. 23883-BY, Washington DC.

LARGE-SCALE DISASTERS – ISBN 92-64-02018-7 – © OECD 2004

Key issues

3. Impact on insurance and financial markets

by Sebastian Schich

The financial system and real activity are closely intertwined. When financial systems perform badly, they can hold back real activity, while, when they perform well, they tend to facilitate faster growth.[1] But the causality also runs in the other direction. Financial market prices and trading volumes react to shocks affecting the real economy such as large-scale disasters, as these events have an impact on the value of existing assets.

Impact on financial and insurance markets

The financial market reaction in the case of the 11 September attacks was much more pronounced than in the case of some other large-scale disasters, judged by developments in equity market indices following the five largest catastrophes in terms of insured losses.[2] In the case of the 11 September attacks, the physical infrastructure of a major financial centre was directly affected. Following these attacks, equity market indices for all G-7 countries, with the exception of Japan and world-wide indices, all recorded negative excess returns. Negative market responses were often much less pronounced and limited to fewer countries in the case of other large-scale disasters. For example, winter storm *Daria* affecting several European countries in January 1990 did not seem to have had a noticeable adverse effect on equity market prices. Hurricane Andrew in August 1992 had an adverse effect on broad US equity market indices, but not on broad world or European-wide indices.

Among different sectors, market participants expected the insurance industry to be particularly hard hit by the fallout from the 11 September attacks. As a large amount of losses associated with the attacks was insured, insurance sector equity market indices recorded much larger declines than the broader indices in most markets. It is likely that this reflects the relatively large amount of *insured* losses.[3] A notable feature of the attacks was the accumulation of (insured) losses across a wide variety of insurance contracts: property, liability, business interruption, life, accident and health. While the final total costs for the insurance industry of the 11 September attacks are yet not precisely known, they are estimated to be between USD 40 and 50 billion, suggesting a need for recapitalisation.

As a general rule, when large-scale disasters deplete capacity in the industry, the subsequent pricing power apparent in premium hikes following such episodes encourages the entry of new capital in the insurance industry. This was what happened following the catastrophic events in the early 1990s.

Specifically, in response to the apparent lack of capital and the high premiums for catastrophe reinsurance in the mid-1990s after hurricane Andrew and the Northridge earthquake, there was a large influx of capital, especially for catastrophe reinsurers. As a result, the capitalisation of the property-liability insurance in the United States steadily increased during the second half of the 1990s. However, more recently, less capital than might have been expected has been drawn into the sector, with a reported net figure of USD45 billion, a figure that should be seen against the background of an estimated loss in capacity of USD200 billion for the industry world-wide as a result of high liabilities, equity market declines and asset impairment more generally. So far, no major bankruptcies have occurred in the insurance industry in spite of the magnitude of the payments associated with the 11 September attacks and the losses associated with declining equity markets and asset impairment more generally. However, it has been suggested that the insurance industry may not be able to withstand another shock of a similar magnitude.

The 11 September attacks, as well as some earlier large-scale disasters, highlight the crucial role that a well-capitalised insurance sector plays for the proper functioning of the economy, as the economy's ability to reallocate risks through insurance is an essential precondition for its functioning. Insurance is a mechanism by which investors or consumers can transfer risk to other parties more capable of bearing it and thus can engage in activities that they otherwise would not if exposed to the risk. But if insurance cover is limited or available only at high prices, economic activity is adversely affected. Typically, such adverse effects could work through at least two channels. First, the lack of insurance prevents some economic activity from going forward. Indeed, a typical reaction of the insurance market following new catastrophes of unexpected magnitudes often consists of the initial withdrawal of insurance cover for the concerned risk class. Second, when new risks are priced, the higher cost of insurance cover makes economic activity more costly, possibly even resulting in misallocation of scarce resources.[4]

How the shrinkage of insurance coverage exactly feeds through to economic activity and to which sectors of the economy depends on the type of insurance policies affected. Unlike previous significant catastrophes, where natural hazards mainly affected (personal) property, the terrorist attacks of September 11 had an effect on a variety of insurance classes, such as property damage, aviation liability, business interruption and life liability. As a result, many sectors have been affected and anecdotal evidence suggests that the shrinkage of insurance coverage had the strongest impact on aviation and transport, as well as on numerous other sectors such as manufacturing, energy, real estate and construction. For example, the debt financing for many real estate deals typically includes a stipulation that the project must carry sufficient insurance to protect the lender. If the borrower is unable to secure sufficient coverage, lenders may be unwilling to make the financing available. Indeed, according to the Bond Market Association, an amount equivalent to about 10 per cent of the commercial mortgage-backed securities market has been suspended or cancelled due to issues of terrorism insurance in the United States following the 11 September attacks. The commercial property market was significantly affected as commercial property and liability insurance rates not only received steep increases, but also became completely unavailable for "target" structures such as chemical and power plants and "iconic" office buildings. As a result of the lack of insurance coverage, construction of commercial properties dropped dramatically in the United States, as has been pointed out by the President of the United States.[5]

Challenges for markets and policy implications

Market behaviour

It is not clear whether large-scale disasters have a lasting effect on financial market behaviour. Effects on financial market prices and volumes beyond the short term are difficult to identify because these variables are influenced by a large number of different factors in a complex way. While it cannot be excluded that there are lingering effects of exogenous shocks such as large-scale disasters, one would expect them to decay over time. Typically, one would expect the effect of exogenous shocks on market behaviour to be more limited than that of shocks that are generated within the system, such as the stock market crash of 1987 or the LTCM crisis of 1998. Indeed, it appears that financial markets have continued to operate properly beyond the very short term period following disasters. For example, financial markets have generally played a useful role in channelling funds between different countries during the later reconstruction period. This facilitates the fiscal reaction to a large-scale disaster in the country affected and allows it to have recourse to international savings when domestic savings are insufficient to fund domestic investments, thus facilitating consumption smoothing over time. By contrast, in the case of the insurance market, a change in market behaviour following large-scale disasters is more noticeable. One important channel through which this generally occurs is the reduction in insurance capacity that is associated with the payment of claims for insured losses. To mitigate this effect, financial resources to withstand such shocks could be increased through greater involvement of the world-wide capital markets (see also companion note by DAF).

Market infrastructure

It appears that changes are occurring in the financial market infrastructure, as market participants attempt to reduce the vulnerability of that infrastructure with respect to large-scale disasters and especially with respect to targeted terrorist attacks. Large-scale disasters highlight the risks associated with the *concentration* of the financial services industry in limited geographical areas that could be observed during the last decades. This has at least two aspects. First, the offices and staff of financial service providers, such as commercial and investment banks, insurers, brokers, dealers and exchanges are in many countries located in a limited area, often mainly in one city in the country. A high concentration of offices and staff in a limited area increases the vulnerability of the financial system to physical damages in that area. Second, because firms are physically concentrated in the same area, they generally tend to rely on the same telecommunications and information technology networks. Consequently, disruptions of those networks tend to affect a large number or even all participants. Against this background, the transaction cost advantages of concentration must be weighed against the disadvantage of greater vulnerability to (physical) shocks.

Policy measures

Crisis management plays a key role in helping to restore confidence and safeguard the financial system. In the case of the 11 September attacks, decisions taken by the Federal Reserve, other central banks and governments were essential in this respect, including the injection of large amounts of liquidity into the system, as well as regulatory forbearance and tailored support. As a result of all these policy actions, financial markets and institutions were able to continue to operate without any major disruption (see Box 2.2). Market participants seem to have viewed policy actions as timely and appropriate and the financial market response to the event was less than might otherwise have been expected.

Thus, one important policy lesson to be drawn is that the financial market reaction depends, among other things, on policy reaction.[6] The capacity of policy makers to react *swiftly* seems to be important, as the speed of the response was seen as key in restoring market confidence, contributing to a feeling among market participants that the pressures they faced were temporary. To enhance the capacity of financial authorities to react swiftly to sudden shocks, *contingency planning* by financial authorities is required. Contingency plans which specify how central banks can inject liquidity into the banking system have the potential to enhance the speed of such action and thus either prevent liquidity disruptions or at least mitigate their effects on the payment system. Contingency planning specifying how to adapt regulations or supervisory requirements in the case of an unusual and large-scale disaster, could speed up responses as well as facilitating the maintaining of a level-playing field for private market actors. One important issue in the context of any contingency plan set up by public authorities is to what extent such plans should be made transparent beforehand.

Box 2.2. **A case study: Financial markets and emergency policy responses to the 11 September attacks**

The physical destruction resulting from the 11 September terrorist attacks deeply affected financial activities that were largely concentrated, together with the supporting information-technology and communication infrastructure, in Lower Manhattan. A very large number of offices and office equipment, as well as telephone lines and data communications lines were destroyed, with adverse effects on the payment systems. Although the core payment system continued to work, telecommunication and operational failures among major financial institutions led to significant liquidity bottlenecks for several days, as many firms were unable to meet their daily payment obligations through normal market arrangements. Several markets including the large and typically relatively liquid one for US government securities were affected. Many markets world-wide, even if not directly experiencing any major technical problems, were characterised by a "flight to quality", widening risk premia, increasing volatility and impairment of liquidity. Particularly noteworthy was the effect on a financial market segment where capital and insurance markets meet, *i.e.* the market for catastrophe bonds, in which there was a very large change in required returns after the attacks.

To mitigate the adverse effects on banks and prevent a significant disruption in the payment system in its jurisdiction, the Federal Reserve used its normal open market operations to inject record-setting levels of liquidity into the financial system during the days following 11 September.* Furthermore, the Federal Reserve also arranged large currency swaps with other central banks, which in turn lent US dollars to financial institutions in their jurisdictions that were in need of dollar liquidity. In the United States, various prudential requirements were eased in order to prevent and/or alleviate credit and liquidity disruptions and allow participants to complete transactions regardless of the sometimes significant, though temporary, balance sheet distortions that these might generate. In a number of jurisdictions, particularly in Europe, insurance supervisors amended or waived temporarily provisioning or accounting rules to prevent such rules from forcing insurance companies to sell equity in a declining market. Emergency measures were also taken by regulators and exchanges in the securities sector. While regulatory responses have differed across countries, intense communication has taken place. In the United States, equity markets were closed subsequently to the attacks for four days and most bond trading, including government securities trading, halted for two days. As well, trading in US securities on European markets was suspended and only resumed a week later.

* The discussion in this paragraph draws on "Issues Paper: Potential supervisory implications of 11 September", by the *Task Force on Extreme Events* of the *Joint Forum*, 19 August 2002.

Intense communication and co-operation between different policy, regulatory and supervisory authorities have facilitated crisis management in the case of the 11 September attacks. It is likely that the benefits to be derived from this type of international co-operation are the higher, the larger is the material impact associated with a disaster compared to the size of the economy and the more other countries are affected, either directly or indirectly. As discussed above, while the 11 September attacks targeted a specific area in the United States, the direct and indirect impact on financial and insurance markets was more widespread. Looking ahead, the benefits of international co-operation could be expected to be even higher if a disaster occurred that would have an even greater dimension than the 11 September attacks.

The 11 September attacks have highlighted the benefits to be had from an private-public communication and co-operation in at least three aspects. First, there are constraints on the types of risks that can be insured, as well as on the magnitude of potential losses than can be transferred onto private insurance markets, which suggests some form of private-public co-operation. Second, intense communication between public authorities and private financial and insurance market participants has facilitated crisis management. Third, providing information about policy action to market participants has helped to reduce uncertainty, and thus to lower risk premia and enhance market liquidity. Much of this has occurred on an *ad hoc* basis, so that an important question is to what extent such communication and co-operation should be formalised *ex ante*.

Notes

1. See *e.g.* Leahy, M., S. Schich, G. Wehinger, F. Pelgrin and T. Thorgeirsson, "Contributions of Financial Systems to Growth in OECD Countries", OECD Economics Department Working Paper No. 280, 2001.

2. A list of the largest most costly large-scale disasters in terms of losses insured is made available by Sigma. See Swiss Re, Sigma, No. 2/2003, "Natural catastrophes and man-made disasters in 2002: high flood loss burden".

3. In this context, note that in the case of the Kobe earthquake in Japan in 1995, the proportion of insured losses was relatively low and that the event figures only 10th among the major large-scale disasters in terms of losses insured. It is feasible that the relative share of insured as compared to non-insured losses has implications for relative equity market valuations, as the latter may reflect expectations about burden sharing.

4. See for a discussion "The Economic Consequences of Terrorism", *OECD Economic Outlook* 2002/1, June 2002.

5. "President Calls on Senate to Act on Terrorism Insurance Legislation", Remarks by the President to Business Leaders, White House press release, 8 April 2002.

6. These issues are discussed by the *Task Force on Extreme Events* of the *Joint Forum*.

Key issues

4. An insurance perspective on terrorism and other disaster management

by Cécile Vignial-Denain

Since 11 September 2001, both governments and the private sector, including in particular insurers and reinsurers, have initiated a far-reaching reflection on the ways to cope with the threat of future catastrophes of comparable or even greater magnitude. To this end, drawing lessons from the management of other types of past disasters and sharing related information and experience at an international level appeared highly instructive.

Managing terrorism and other large-scale disasters: an increasingly challenging task calling for a strong involvement of the insurance sector

Although terrorism risk is characterized by certain unique features, it raises challenges similar to those posed by other types of large-scale catastrophes, i.e.

- a financial challenge: the compensation/reparation process after large scale disasters requires very large financial resources; (this issue will be dealt with in the chapter on compensation);
- an operational challenge: large-scale disasters translate into a convergence of innumerable claims[1] that need to be filed, assessed, and compensated within a short time frame.

In both respects, the insurance sector can provide a key contribution.[2] Insurers and reinsurers are main players in the settlement of catastrophes. Their experience and expertise in risk assessment, claim management and compensation is well-known, while they increasingly play a crucial although less advertised role in the area of prevention. Besides, insurers may also be involved in the provision of emergency assistance. Lastly, the role of insurers tends to expand as a consequence of increasing awareness of risks and of the need to better protect populations against expanding risks. An obvious example of this trend is the recent introduction of a requirement to purchase insurance for certain types of large-scale disasters, which forced insurers to cover risks sometimes considered as uninsurable in the past.[3]

Nevertheless, the very characteristics of large-scale disasters – potential extreme magnitude, difficulties in assessing and pricing risks, etc. – raise a number of complex technical issues for insurers. Major catastrophes are often on the borderline of technical insurability. Insuring against disasters is all the more difficult in that the characteristics and nature of these risks are evolving fast.[4] Consequently, exchange of information on processes and techniques tested around the world to manage catastrophes becomes extremely important to optimise the role of the insurance sector therein.

Lessons to be drawn from the management of past disasters

Lessons to be drawn from the management of past catastrophes could be outlined through the following checklist.

Ex ante measures

The sophistication of insurers' risk forecasting techniques and preventive strategies will help them to fine-tune their pricing and compensation abilities, reducing the potential impact of disasters and sometimes minimizing the probability of catastrophe occurrence (for industrial pollution risks for instance).

Improved risk assessment

Sufficient information on probability and magnitude of risks needs to be gathered in order to adequately price insurance contracts and to enable the constitution of sufficient provisions. The long experience of insurance experts in the evaluation/modeling of risks and in scenario building is a key asset in this respect. Although large-scale disasters as a whole are obviously more difficult to price than motor or other mass insurance risks, series of data on past natural catastrophes for instance are available which allow for estimation on trends and potential losses from future catastrophes. With new computer modelling techniques, based on geographical information systems (GIS) and hazard analysis, the available data can be enriched to improve claims forecasting. The major insurance and reinsurance companies have developed such state-of-the-art risk assessment, rating tools, and hazard maps for earthquakes, tropical cyclones, winter storms and floods. For terrorism risks, in the absence of a historical series of terrorism acts comparable in nature to those of 11 September, insurers do not dispose of truly relevant information on probability and magnitude of risks, and modelling the occurrences of attacks appears as a highly complex exercise. Nevertheless, insurers endeavour to reassess their risk profile.[5] Coordination with highly specialized experts outside of the insurance sector, e.g. experts from intelligence agencies (in the same way as meteorological experts are consulted in regions exposed to specific natural disasters), has also proved helpful.

Risk assessment and modeling is obviously a crucial phase for which research and investment should be further promoted,[6] especially in the area of terrorism risk where the WTC attacks have revealed the industry's unpreparedness. The sharing of information and research, and of risk assessment and modeling techniques, between companies and between private and public actors, including the academic community, should also be encouraged.

Preventive strategies

Preventing the occurrence of disasters/reducing their potential impact:

- Insurers play an increasing role in the "remodeling" of risks, in order to limit their client's exposure and their own. "In modern environmental insurance, professional risk-carriers have the knowledge and technical ability needed to actively intervene on the risk features during a new phase, which can be named: risk remodeling",[7] before the actual transfer of risk. This requires a careful evaluation and classification of the risk, comprehensive inspection of the property insured, evaluation of the adequacy of safety measures, protection systems and emergency plans, etc. Once the risk has been properly assessed, the risk carrier will cooperate with the prospective insured in order to reduce the risk and to enhance loss prevention strategies.

- To complement risk remodeling strategies, premiums and deductible levels should be closely tied to the effective prevention efforts, to avoid moral hazard.

- The role of advice and education of policyholders by insurers should be further encouraged. Potential policyholders should also be made aware of the risks associated with insufficient insurance cover or lack of cover.

- Contingency planning and co-ordination among all those in charge of disaster relief should be organised ex-ante, and early warning systems should be further developed.

Preventing the disorganisation of insurance companies' claim settlement services:

To avoid disorganisation when a large amount of claims are lodged in a short period of time, the following precautions should be taken:

- design ex ante crises management structures;

- train people accordingly, allocate a precise role to key management in case of crisis;

- address problems like information system back-up or the possible need for the insurance company to operate from a different location, if the insurance company offices and operation have been affected by the disaster.

Ex-post *management measures*[8]

The design of emergency procedures is a key step to enabling an efficient treatment of requests for assistance and claim flows.

Non-financial crisis management

- Special arrangements should be designed, as particular events occur,[9] or under the form of a global standing body,[10] to help the insurance industry to co-ordinate its actions to ensure that the people affected by a disaster get rapid and effective support.

- Operational procedures should be defined,[11] e.g. setting up of a crisis task force – gathering representatives of emergency services (fire-fighters, doctors, etc.), of the insurance industry, and of the government –,[12] a toll-free telephone number,[13] special communication procedures, emergency assistance for victims for which insurers can be called to play a role (typically for short term accommodation assistance, or for the granting of additional living expenses for people forced out of homes).

- Post disaster reviews of performance should also be systematically conducted.

Claim management

Monitoring the settlement procedures[14]

- Operational procedures need to be defined, such as a special claims filing procedure (e.g. a common claims form for use by all companies)[15] or a special time limit for filing claims.[16]

- To make the compensation faster, certain exceptional settlement measures could be considered. In respect of property damage for instance, loss handling could be subject to a prioritization, applying the philosophy that "the worst comes first",[17] compensation could be granted without deductibles or depreciation coefficient, repairs could be made without appraisals for damage up to a certain amount,[18] and special rules could be established for pay-outs (e.g., with companies agreeing to pay advances[19] of a specified standard amount).

- To ensure fairness and equity in treatment in respect of bodily injury, all examinations necessary to assess compensation amounts/appealing settlement amounts could be managed by a single body.

Differentiating between the treatment of property damage and bodily injury

Settlement procedures may need to take longer for bodily injury than for property damage. Also, claims in respect of bodily injury could be given priority, especially when the law imposes ceilings on cover.

Monitoring insurance fraud

Insurance fraud, which may develop because of the rapidity with which claims have to be settled, should be monitored and severely punished.

The role of governments

The role of the insurance sector in the management of large scale disasters should be supported by an adequate regulatory framework. Government action in this respect could in particular involve the efficient protection of the population at risk and the reduction of damages by means of mitigation regulations. Adequate legislation regarding for instance building and infrastructure safety or land use management and urban development[20] should be enforced, while tax and subsidy systems applicable in the insurance area should provide appropriate incentives for risk coverage, prevention and for the reduction of moral hazard.

More generally, governments should highlight the threat entailed by large-scale disasters and enhance risk awareness, while promoting an insurance culture among populations.

Notes

1. 78 000 claims were lodged after the AZF plant explosion in France for instance.

2. It is precisely to cope with unpredictable large losses, originally in the maritime transport area, that insurance was created.

3. Natural catastrophes were traditionally considered as uninsurable in France until the 1982 law, which made insurance cover against such perils compulsory.

4. The 11 September attacks highlighted the change in the nature of terrorism risk. Such risk, considered until that date as marginal and generally covered at no additional cost, turned out to be an incommensurable threat on all nations necessitating important changes in insurers' risk management techniques.

5. To this end, insurers may also seek the assistance of specialised firms, such as Eqecat, RMS (Risk Management Solutions) or AIR, which have developed for the United States new models based on extensive data related to possible exposure to terrorism risk in this country.

6. This could be done at company or sector level: pollution insurance pools established in various countries not only to aggregate capacity, but also to develop new products and share information, statistics, and experience, is an experiment that could be extended to other countries/insurance lines.

7. *Environmental Risks and Insurance: A Comparative Analysis of the Role of Insurance in the Management of Environment-related Risks*, Pr. A. Monti, OECD, 2003.

8. Some of these measures may be taken by insurers on a contractual basis, while others may need to be confirmed by legislative acts. In France, the 30 July 2003 law No. 2003-699 30 relative to the prevention of technological and natural disasters is a codification of lessons drawn from the AZF disaster.

9. E.g. a number of operational procedures were defined and stipulated in a formal agreement after the explosion of the AZF plant in Toulouse – the largest industrial disaster ever to have occurred in France. A special committee was created convening at regular intervals all the parties concerned by management of the

claims (insurers, representatives of the victims, Total Final Elf, the prefecture, the city of Toulouse, lawyers, etc.).

10. Like the Insurance Disaster Response Organisation or IDRO in Australia, created in March 2000: IDRO's primary mission is to help the insurance industry to co-ordinate its actions with government agencies and emergency services to ensure that the people affected by a disaster get rapid and effective support and to provide them with a single point of contact. IDRO is essentially a partnership of insurers, re-insurers, insurance brokers, and loss-adjusters. In Canada, the Insurance Bureau of Canada's Claim Emergency Response Plan, created following the 1979 Mississauga rail-crisis and the violent hail storm that struck Calgary in 1981 answers to the very same objectives.

11. These may depend on the level of emergency. Emergency levels have been rated in three categories in Canada, depending on the seriousness of the damage, to allow the fine-tuning of emergency procedures accordingly.

12. The ACT Bushfire recovery Taskforce had for instance been set up after the major forest fires that ravaged the Canberra region in January 2003. It was financed on public funds.

13. In the wake of the power outage of 14-15 August 2003 in the United States, the New York Insurance Dept activated a toll-free hotline to assist New Yorkers whose property and businesses were affected by the power failure. *Impact of the recent power blackout and hurricane Isabel on the financial sector,* Office of public affairs, October 20, 2003.

14. In the Netherlands, the insurers' association has drawn up a disaster scenario which contains *inter alia* rules concerning the joint management of claims processes by all insurers concerned.

15. Many operational initiatives can help policyholders with claims lodging, such as the mobile "Help van" deployed by one large Australian insurer after the Canberra bush fires to assist policyholders with additional claim support and immediate claim assessment.

16. In France, insurers have to compensate policyholders within three months in case of natural disasters. Three weeks after the Canberra Bushfires, insurers had already paid out 54 millions dollars in claims.

17. As suggested by the Claim Emergency Response Plan in Canada.

18. Such measures were for instance enforced in France after the AZF explosion.

19. On December 8th, several insurers had already committed to pay advances after the early December 2003 floods in South West France, even before the Government had issued the decrees on natural catastrophe situation. *Les Échos,* 8 December 2003.

20. In Turkey, strict standards have been imposed on new construction work since 2000 to minimise potential damage from earthquakes. Similarly, in France, the National Fund for Prevention of Major Natural Catastrophes created by the 2.2.1995 law, allows the expropriation of properties threatened by natural perils endangering human lives.

Key issues

5. Compensation issues

by Cécile Vignial-Denain

Who should compensate victims of large-scale disasters? Who can indemnify for extreme loss amounts at short notice?

Major catastrophes raise a critical financial challenge. This concern is all the more worrying in that the magnitude and the impact of such disasters have increased markedly over the past 30 years. Natural catastrophes are an example of a rapidly evolving threat: due to a conjunction of factors, including global warming and urban development in areas exposed to unfavourable natural conditions, insured losses from natural disasters have increased 15-fold since 1960. Similarly, the 11 September attacks, which entailed losses of about USD 40 billion according to low estimates, are an obvious illustration of the change in the magnitude of terrorism risk, and have raised the spectre of possible future "mega terrorism" attacks.

Compensation for terrorism and other large scale disasters: a growing concern for the insurance industry and the state[1]

Cover against such disasters can be sought from three main parties: the (re)insurance industry, governments, and the financial markets. Their respective contribution is however subject to various constraints.

Can the private sector cope?

The insurance, and ultimately the reinsurance sector support most of the costs of large-scale disasters. However, insurability is not infinite: technical and financial criteria define its boundaries.

Private insurance capacity is limited

The capacity of national insurance markets, even backed by international reinsurance, is finite,[2] and the insurance sector may therefore not be able to cover any compensation requests without endangering its financial stability. In 2001, the total capital and reserves of the world's non-life insurance and reinsurance industry were estimated to be about USD 500 billion, of which only about USD 130 billion can be considered available to cover large potential commercial risks.[3] Hence the September 11th losses were within the capacity of the insurance industry but another similar sized loss would severely test its financial resources. Reinsurers would be the most threatened, as they accept the largest losses, often on a non-proportional risk sharing basis. For instance, they will eventually bear an estimated 70% of the total insured loss amount entailed by the 11 September attacks.

Insurance of large-scale disasters suffers inter alia from difficulties encountered in the mutualisation of risk exposure. Natural catastrophes risks, and to some extent terrorism risks, fail to meet the mutuality condition which allows the law of large numbers to apply efficiently. Not only are losses very large and infrequent, but also risks are highly correlated, and the incidence of loss is often concentrated in particular geographic areas or aimed at high profile targets, entailing an important anti-selection phenomenon and limiting the community of insured and the premium base.

Private cover may not always be available

Faced with fast-evolving disaster risks, hardly predictable in their nature, magnitude and frequency, insurers may decide to adopt a prudent stand and, at least temporarily, exclude them altogether from their cover as uninsurable. This was for instance the reaction of most reinsurers, followed by the insurance industry, after the heavy losses incurred on 11 September, which resulted in a drastic capacity shortage.

Private cover may not always be affordable

Affordability is a key issue for the insurance against earthquake, flood or other natural catastrophe risks, as for the insurance of terrorism risks. Due to lack of sufficient information on the characteristics of future disasters needed to calculate an "actuarially fair premium", insurance prices tend to rise considerably. This in turn may discourage consumers to seek insurance cover, which may simply be unaffordable for some of them. Such phenomenon was observed after the WTC attacks, when the drastic restriction of cover was accompanied by a very substantial rise in the level of premium.

The penetration of large-scale disaster insurance is often limited

Unaffordable premiums, anti-selection, but also the lack of awareness of the risks incurred limit insurance penetration. Only part of losses is therefore borne by the insurance sector after a disaster. Because of the low market penetration and strict limits on flood cover in certain regions, the insurance industry faced a total insured loss estimated at USD 3 billion[4] after the summer 2002 floods in Europe – only about 10-20% of global economic losses.[5] These gaps in insurance cover entail substantial negative economic effects.

Is State backing a relevant solution?

There is a clear rationale for government involvement in the compensation of large-scale disasters: the potential magnitude of risks and their technical characteristics which may not match insurability criteria may result in market failure.[6] Besides, political and social consequences of large scale disasters as well as the threat of major disruption in key economic sectors and related spill-over effects may lead governments to intervene. Lastly, it should be underlined that State backing has often proven to be a key condition of private player's involvement above certain levels of risk exposure.

Such state involvement can take several forms, from regulation promoting the development of private coverage to the provision of reinsurance of last resort, loans, or free compensation of losses.

While States are usually closely involved in the compensation process after a disaster for the above-mentioned reasons, the setting up of longer-term arrangements based on government support may also appeared justified under certain conditions. It however raises a number of issues that should not be underestimated. In particular, the fear of possible "crowding out effects" for the private sector that could discourage the adaptation of insurance markets is a strong deterrent. Also, direct compensation of certain types of disasters or losses implies complex budgetary trade-off and possible competition distortion. Several OECD countries have therefore opted for a temporary state involvement, which usefulness is regularly reassessed.

Financial markets: a viable alternative?

Recourse to the financial market is another potential solution to capacity deficit. Given the concern about the enormous and increasing financial losses caused by large-scale disasters in recent years, and their possible consequences on insurers' and states' finances, the idea of transferring some of these risks to capital markets – representing USD 29 trillion at the end of the first quarter of 2003 – raised growing interest.[7] As demonstrated in Figure 1, financial markets can contribute to extend financial capacity through many channels. Besides, they have a particularly interesting role in the specific case of catastrophic risks as an alternative to traditional insurance markets via the issuance of catastrophic bonds.[8] The market for CAT bonds emerged from the times of capacity shortage which followed hurricane Andrew in 1992. The hardening of commercial insurance market and the shrinkage of capacity due to the fall in financial markets in the recent past and to the consequences of the WTC attacks were seen as a promising environment for such insurance-linked securities. After 11 September, it was also thought that terrorism bonds could be issued along the same lines to solve terrorism insurance capacity shortage.

However, contrary to expectations, CAT bond issuance has reached a plateau and now remains rather flat with an average annual issuance volume of about USD 1 billion since 1997, while CAT bond prices remain somewhat higher than conventional reinsurance products. Regarding terrorism, terrorism risk securitisation has not had the kick-start anticipated from the hardening of the conventional market, and many obstacles still prevent its development. In particular, the structuring of such securities is still expensive, while

Figure 1. **Risk-sharing networks**

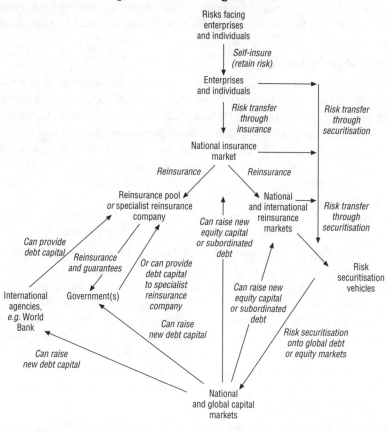

Source: G. Dickinson, Insurability Challenges for Large Terrorism and Natural Catastrophe Risks, unpublished manuscript, OECD, 2002.

investors are reluctant to buy a product, of which the underlying risk is highly complex, as in the case of terrorism. Therefore, the first transfer of terrorism risk to capital markets only occurred in September 2003, with a USD 260m of "cancellation bonds" issued by the FIFA to insure against terrorism-related property damage at the occasion of the 2006 Football World Cup.[9] While promising, securitisation of catastrophic risks only provides a marginal solution to traditional market capacity problems for the time being.

Lessons to be drawn from the compensation process of past catastrophes

Although States will continue to be called upon in the compensation process of large-scale disasters, the private (re)insurance sector is more and more solicited to carry part of the financial burden. Private capacity building is essential, and should alleviate the role of the State in the long-run. Meanwhile, private-public partnership through catastrophe compensation schemes appears to be an efficient solution to finance the costs entailed by catastrophic events.

Private capacity building

Insurers have traditionally tended to find innovative solutions to stretch the limits of their capacity when faced with losses of unexpected magnitude. Such endeavours translate into capacity building through capital injection in the concerned branches of activities or through the creation of new ventures.[10] It could also materialise in the setting up of private pools. This type of risk-sharing agreement is a typical answer to limited capacity in the insurance sector, traditionally organised along re- and co-insurance arrangements. In Austria or Russia, private pools have been set up to help in compensating terrorism risks. Similarly, in Switzerland, an entirely private scheme created in the early 50s' is devoted to the compensation of natural catastrophes.

Private capacity building can be efficiently supported by a wide array of legislative measures, initiating for instance tax incentives to large-scale disaster provisioning, specific forbearance measures, changes in accounting of assets, special levies on insurance premiums, or compulsory insurance for specific types of large-scale disasters. Developing insurers' ability to compensate for disasters also entails the curtailing of their liability. For environmental pollution for instance, liability law should be designed in order for the exposure of insurers to be as predictable and manageable as possible; it should in particular be non retro-active and stable to allow a fair degree of confidence of suppliers.

Private-public partnership and the setting up of catastrophe compensation schemes

In many developed countries exposed to specific large-scale disasters, the creation of specific compensation schemes based on a private-public partnership has emerged as a necessity to cope with the financial consequences of large-scale disasters.[11] Such schemes have been motivated by the "need for specific, highly specialised treatment for technical-actuarial, financial and management aspects" of extraordinary risks,[12] most often in the realm of a disaster. The design of such schemes will reflect market specificities, exposure to risks, as well as technical and policy (in particular as far as they reflect the degree of intended State participation) choices, regarding the scope of coverage targeted and the modalities of financing.

Scope of cover

Perils covered: catastrophe schemes are either:

- devoted to one type of disaster: *e.g.* flood compensation scheme in Australia, Denmark, the Netherlands or Poland; or terrorism compensation schemes in all countries where they exist except in Spain; or

- covering a variety of disasters to which the country may be particularly exposed: *e.g.* scheme covering various natural disasters, like in France, Iceland, Japan, Mexico, New Zealand, Norway or Switzerland; schemes covering both natural and technological catastrophes like in Belgium; scheme covering both natural perils and "acts with social repercussions" (terrorism, rebellion, riots, etc.) as the Spanish Consortio. While less flexible, schemes encompassing various types of perils may allow a greater mutualisation of risks;

Geographical coverage: schemes may cover a single region (*e.g.* California or Florida), national states, or several states (the first initiative in this respect is the European Union solidarity fund, created in 2002).

Triggers: in order to avoid as far as possible gaps in coverage or legal prosecutions, triggering events should be defined as unambiguously as possible. Governmental certification of triggering acts or losses entailed, before compensation could be claimed – as for terrorism acts in the US or natural catastrophes in France or Turkey[13] – may help in this respect, but it may also involve delays and political bias.[14]

Type of losses/insurance lines covered: country options – specific focus on infrastructures/farmers/ low-income victims/residential *vs.* commercial properties –, may reflect trade offs necessitated by the limited financial capacity of the scheme, as well as policy choices.

Optional or mandatory insurance: compensation schemes often entail an obligation to purchase insurance, in order to avoid anti-selection, optimize the mutualisation of risks and broaden the insurance penetration and the capacity of the scheme.[15] For instance, France, Iceland, Norway, Spain and Switzerland have all introduced compulsory insurance against natural disasters. Consequently, penetration is very high in these countries (95 to 98 % of homes in France[16]).[17]

Financing of the programme

The role of the State varies substantially between the various national schemes, from targeted schemes like the Mexican FODEN against natural disasters for instance, that is entirely funded and managed by the government, to several compensation mechanisms (in Belgium for natural and technological perils, Spain or the Netherlands for floods) where the State has a subsidiary role and covers risks that are not commercially insured, or to most terrorism compensation schemes where States have a supplementary role

Figure 2. **Multi-layer risk sharing mechanism**

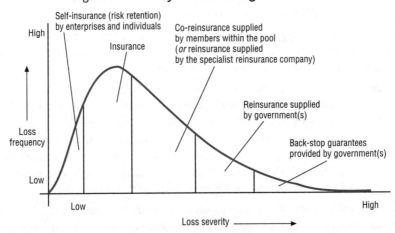

Source: G. Dickinson, *Insurability Challenges for Large Terrorism and Natural Catastrophe Risks*, unpublished paper, OECD, 2002.

and typically intervene above very high amounts of losses or as a last resort reinsurer, adding a top cover to multi-layer risk-sharing mechanisms (involvement of the various stake-holders according to loss frequency and severity in such schemes is illustrated in Figure 2).

Compensation of mega-risks

Mega risks refer to risks exceeding the current financial capacity of the insurance industry and the government of a given country – *i.e.* the industry capacity and the State ability or willingness to indemnify losses without endangering the national economic stability. For instance, scenarios built on the basis of new-style terrorism risks, which could take the form of physical attacks, biological or chemical attacks, or a systematic undermining of computer systems, or a large-scale industrial or technological catastrophe, may by far exceed the compensation capacity of small OECD countries. Adequate financial response for such risks may only be provided via an international mechanism involving States as last resort capacity,[18] and possibly the financial markets in the future.

There is clearly no ideal risk sharing model to cover large-scale disasters, since each scheme should adapt to the specificities of the national market and to pre-determined policy goals. However, decision-makers should consider several underlying principles, which have been *inter alia* highlighted by the working group on UK Pool Re's reform,[19] and mainly:

● separate the short-term needs from long-term challenges;

● allow for flexibility since the characteristics of major disasters may evolve over time, as may insurance techniques and willingness of States to be involved in compensation;

- adopt a balanced approach between the role assigned respectively to the insurance industry, financial markets and the State, to avoid discouraging the private sector from adapting to these risks; and

- properly assess the negative externalities stemming from insufficient large-scale disaster coverage for the rest of the economy.

Notes

1. At the occasion of the Ministerial Meeting held on 15-16 May 2002, OECD ministers recognised the adverse effects of the shrinkage of affordable insurance cover for terrorism risks and mandated the OECD to develop policy analysis and recommendations on how to define and cover terrorism risks and to assess the respective roles of the insurance industry, financial markets and governments, including for the coverage of "mega-terrorism" risks. The OECD Insurance Committee is currently working on the implementation of this mandate. It did as well address the issue of the indemnification of large-scale disasters.

2. This limit is set by the available capital that the global insurance and reinsurance sector holds and can gain quick access to, as well as the pattern of future prices after a large loss.

3. The conclusions of certain other research are somehow more optimistic. "Estimating capacity using insurer financial statement data, we find that the (US) industry could adequately fund a USD 100 billion event" – even if "such an event would cause numerous insolvencies and severely destabilize insurance markets", say David Cummins, Neil Doherty and Anita Lo in *Can insurers pay for the "big one"? Measuring the capacity of the insurance market to respond to catastrophic losses*, Journal of Banking & Finance 26 (2002), 557-583.

4. *Hazard review of the year 2002*, Guy Carpenter, January 30, 2003.

5. *Annual Review : Natural Catastrophes 2002*, Munich Re, 2003.

6. For instance in Sweden, a majority of insurance companies do not cover potentially considerable damages caused by dam breaks, according to the prevailing idea that such cover can only be provided by the State.

7. *Catastrophe insurance risks: status of efforts to securitize national catastrophe and terrorism risk*, GAO, September 2003.

8. CAT bonds enable insurers to transfer catastrophic risks to capital markets through a bond issue. "In a typical CAT bond structure, an insurer issues an interest-bearing bond to the capital markets. The payments received are turned over to a special purpose vehicle (SPV), which then issues a conventional reinsurance policy to the insurer. Typically, the funds held by the SPV are given to the insurer if the bond's specified event is triggered, the interest payments usually cease at this point." *The picture of ART*, Sigma No. 1/2003, Swiss Re, 2003.

9. *Risky business, The Economist*, 4 December 2003.

10. Within six months of the 11th September tragedy, several private operators were ready to come back on the insurance market and take advantage of a potentially highly profitable niche. A further USD 20 billion of new capital was attracted into the non-life insurance sector, with USD 10 billion in Bermuda, from new company formations or capital raised by existing insurers and reinsurers.

11. About one half of OECD countries have established compensation mechanisms devoted to major catastrophes. Some countries have several schemes to cover different types of disasters. The main large-scale disaster compensation schemes in OECD countries are the following:

Australia: Federal flood recovery fund (2000), Commonwealth flood assistance package for central and Northern NSW and Southern Queenland, Natural disaster relief arrangement; Belgium: Natural disaster and major technological risks compensation fund (Fond des calamités naturelles et des risques technologiques majeurs) (1990); Denmark: Storm council (stormradet) (1991); France: National disaster compensation scheme (CAT NAT) (1982); National guarantee fund for disasters affection agriculture (Fonds national de garantie des calamités agricoles) (1964); Iceland: Incelandic emergency insurance organisation (1975); Japan: Japanese earthquake reinsurance Co Ltd (JER) (1966); Mexico: National fund for natural disasters (FONDEN) (1996); Netherlands: the calamities compensation act (WTS) (1998); New Zealand: earthquake commission (EQC) (1994), Natural disaster funds; Norway: National fund for natural disaster assistance (Statens naturskadefond) (1961), Norwegian pool of natural perils (Norsk naturskadepool) (1980); Poland: National programme for restoration and modernisation; Spain: Consorcio de compensacion de seguros (1954); Switzerland: Swiss Elemental Pool (1953); Turkey: Turkish catastrophe insurance pool (2000); United States: National flood insurance programme (1968), California earthquake authority (1996), Florida hurricane catastrophe fund (1993), Hawaii hurricane relief fund (1993); European Union: European union solidarity fund (2002).

12. The Consortio de Compensacion de Seguros, Economic Ministry, Spain.

13. Except for earthquakes.

14. These reasons were invoked in Spain to abandon the certification requirement.

15. As highlighted in former OECD studies (see in particular M.Fontaine and H.de Rodes: Mandatory insurance in OECD countries, in *Insurance regulation and supervision in OECD countries*, 1997) , governments typically require the purchase of insurance with respect to three types of potential loss-causing activities: those whose severity could be particularly great with a large number of innocent persons being harmed because of a single event, those whose frequency is sufficiently great to affect large numbers of innocent people independently, and those that are judged to be inherently dangerous. The first case covers large-scale disasters.

16. According to the French national statistics institute INSEE (1999).

17. The fact that insurance is compulsory entails an obligation for the insured to be covered against certain types of risks, but it does not systematically entail an obligation for insurers to join existing compensation schemes. However, when the insurer joins a pool or scheme, it is usually required that his entire portfolio of risks in the area concerned (flood, terrorism, ...) is included in the pooling mechanism to avoid anti-selection.

18. A comparable scheme has already been created for losses resulting from the peaceful use of nuclear energy, which can be compensated through the mechanism foreseen by the Paris convention and Brussels supplementary convention.

19. Pool Re is the mutual insurance company created in 1993 in the UK for the compensation of terrorism-related losses.

Key issues

6. Housing and reconstruction

by Josef Konvitz

Housing is usually the largest category of property loss in a natural disaster. Where and how the housing stock is rebuilt has significant implications for community reconstruction. There is no strict, linear correlation between the magnitude of housing losses and the recovery process. A tight local housing and labour market at the time of a disaster obviously makes re-housing more difficult: because the vacancy rate in the greater Los Angeles area was relatively accommodating in 1999, the effects of the Northridge earthquake were absorbed relatively easily. The short-term recovery of economy activity, even in wartime, is often more rapid than the rebuilding of the housing stock or the reconstruction of a community. The rebuilding process may leave a region with improved infrastructure, housing and other assets, or leave it handicapped, its economy under-performing, with social and environmental stresses. Policy issues related to financial and regulatory instruments, the roles of different levels of government, and the role of the private sector have an impact on housing and community reconstruction.

Housing and reconstruction needs

The scale of a disaster matters: policy measures suitable for a small-scale disaster may be inappropriate or even counter-productive when the losses are significant and when strategic questions are raised about what should be rebuilt, and where. But basic data for analysis is lacking: the US Federal Emergency Management Agency stated in 2001 that no disaster in the United States had been properly costed, including property and business loss, the impact on the local tax base, and unemployment.

The response to a disaster at territorial level, be it flood, earthquake, violent storm, or industrial accident, is invariably an impulse to help. Collective solidarity is expressed at the international, national, regional and local levels: government leaders and their administrations are expected to be visible and useful. Mayor Giuliani's actions in New York were a model for mayors worldwide. A disaster often reminds people of what they have in common. Voluntary efforts, whether through non-governmental agencies and associations or spontaneous, self-organised units, not only provide needed assistance, often with less red tape, but also provide an outlet for people who want to make a direct contribution of service. Worldwide, 150 million people are affected by a disaster each year. The Red Cross helped 10 million in 1990, and over 40 million in 1999.

One expression of collective solidarity is a desire to use the reconstruction process to improve a territory and to reduce the scale of damage in the event of any future disaster. The reconstruction of a devastated area begins almost immediately after the natural or man-made event has occurred. Emergency relief and reconstruction usually involve different authorities and priorities, but occur simultaneously at least for several weeks. Decisions taken immediately after a disaster affect the reconstruction process for months or even years to come.

Although there are strong similarities between natural disasters and a terrorist attack in terms of territorial impact, there are some striking differences. Housing is more likely to be built in disregard of engineering norms and planning constraints (and hence may be more vulnerable to destruction in a natural disaster) than major office or public service buildings, which are more likely to be the target of terrorists. (In the United States, an average of 3 500 people die yearly in housing fires, but only 150 in non-residential fires.) The dependency of urban centres on highly efficient transport, energy and communications systems delivered through complex networked systems is thought to make them vulnerable to a terrorist attack, on the grounds that the disabling or destruction of such systems creates disorganisation on a vast scale. But the evidence of aerial bombardment in World War II as well as some post-war disasters (heavy storms, floods and earthquakes) shows that urban infrastructure systems are more difficult to destroy and easier to repair than is often thought to be the case, that major users can adapt and may have back-up plans, and that workers improvise when either housing or places of business are inaccessible or destroyed. *There is little research concerning the factors that make communities more resilient.* The historic record however may be a poor predictor: San Francisco, Istanbul and Tokyo were considerably smaller when last destroyed by earthquakes than they are now. Equally, it is conceivable that a major terrorist attack could render large numbers of dwelling units unsafe temporarily or permanently, which could well overwhelm the institutional and financial capacity of a country to cope, especially if the alternative were to build a new city.

The importance of communication cannot be over-emphasized. A large-scale disaster, by definition, removes familiar landmarks ("place annihilation"); the experience of the disaster is therefore disorienting. The personal demand for communication during the acute phase of a disaster has been demonstrated both in New York (11 September 2001) and in Turkey (Izmit earthquake). Although it is critical that the authorities provide information promptly and continually, much that people want to know concerns their friends and family. The disaster is lived and remembered in personal terms. The history of postwar rebuilding in Europe after 1919 and 1945 shows that several years are needed for the design of appropriate forms of commemoration, memorials, major civic projects and the like.

An exemple: Kobe

On 17 January 1995, the Great Hanshin-Awaji Earthquake struck the southern part of Hyogo Prefecture (Japan) at 5:45 in the morning. This earthquake, with a magnitude of 7.2, was responsible for the deaths of more than 6 400 people, injuries to 44 000, the loss of 105 000 houses and damage to 145 000. Some 300 000 people had to be housed in shelters following the earthquake. Essential services of power, gas, water and telephone were cut off; key road and rail links were destroyed. Natural disasters are a reminder of how-inter-related the complex elements of a modern city are, but also how flexible and adaptable can be in time of crisis. Kobe, the city at the centre of the earthquake, is one of Japan's major industrial and port locations. The disruption of port activities had ripple effects across the Pacific. The reconstruction of Kobe began while emergency relief was being directed toward the victims, while fires were still being put out, and emergency supplies mobilised. All but 10% of the homeless were re-housed within four years, but port operations and essential services were restored, at least partially within weeks.

Delays in the rebuilding of Kobe were minimized, partly because government at all levels recognized the priority (a special task force brought people from different sectoral ministries together under the responsibility of the Prime Minister), and partly because the city and the prefecture had already analysed some of Kobe's deficiencies in infrastructure, such as its dependence on one main elevated highway to serve the port. The replanning of Kobe's highways (to provide multiple routes and capacity for growth) and the creation and re-organisation of its parks (which can also serve as safety refuges in time of disaster) helped the city to realize strategic objectives which had already been identified before the earthquake. The potential value of on-going strategic assessments of a city's strengths and weaknesses whenever a disaster occurs, is supported by the rebuilding of the centre of Manchester following a terrorist bombing on 15 June 1996 with USD 1bn. in economic losses: a public-private partnership that had helped to prepare the city's unsuccessful bid for the Olympics proved invaluable in guiding the reconstruction (Williams, 2000).

The lessons of the disaster also reflected *social issues* which were addressed in the reconstruction effort. For example, many of those who died were elderly who lived alone in small wooden structures. The rebuilt city provided for mixed social housing which brought people of different generations together, with appropriate facilities. However, the replacement of housing stock has left some residents, especially aged people, with problems

66

of adjustment to the design of new buildings, which do not have a cultural or historical identity; design could have been better used to maintain cultural assets and a sense of place. Positively, the adaptive reuse of abandoned industrial and brownfield sites has been a hallmark of contemporary Kobe.

Disaster preparedness has remained a priority, and particularly to give people a sense of security and to enable them to become more self-reliant. The rebuilt city is organised into districts or neighborhoods able to provide decentralised disaster relief, a pattern consistent with "compact city" concepts which are policy objectives in many OECD countries. Thus, the rebuilding of Kobe combined economic objectives for a more competitive city with issues related to ageing, social housing, land use, and risk reduction.

A significant innovation in public co-operation helped relocate people to new districts and new housing, both to achieve a better mix of younger and older people, and to eliminate wooden structures likely to burn. In each district, an Urban Development Council was formed in units of a single block or of several blocks, 97 in all. The Councils organised study meetings and conducted questionnaire surveys, then summed up the comments of residents and drafted plans for urban development. The city government supported the Councils' activities by establishing Reconstruction Consulting Offices, and sending consultants; this assistance was provided to nearly 300 groups by the end of 1999. The consulting service helps residents examine options for housing and land use choices.

Reconstruction compromised by policy and governance failures

A large proportion of the physical damage related to a flood, earthquake, fire or other natural disaster is in housing and related infrastructure (water and sewage systems, transport, power). Hundreds, perhaps thousands of people may have to be relocated temporarily, but *when shelter facilities are inadequate, or when their use stretches from days to months or years, people often evacuate themselves to other localities, perhaps never to return.* When there is a perception that a stricken area remains at risk and when people lack confidence in the enforcement of building codes (Mexico City, following the 1985 earthquake which left 30 000 homeless), people relocate, thus compromising the regeneration of the area where they had previously lived. (This phenomenon of resettlement also occurred in Istanbul out of fear of what might happen in an earthquake. simply in response to the Izmit disaster.) It does happen therefore that a territory rebuilt after a disaster (*e.g.*, parts of South Florida after Hurricane Andrews, 1992; Mexico City) may have greater disparities higher environmental costs, more concentrated poverty, and less cohesion than before, or in other words, has diminished sustainability. These effects are associated with deconcentration from the centre to the periphery , a phenomenon which may yet occur in the New York metropolitan region following the attacks of 11 September 2001.

When a disaster brings to light *weak enforcement or non-compliance with building and planning codes*, thereby increasing the loss of life and of property, the self-organising relief efforts by civil society may be accompanied by strong feelings of distrust for the public authorities. Key issues involving the regulatory process, which include specifying what should be put in codes, what to do to bring older buildings and structures up to date, and enforcement, raise questions about co-operation among different levels of government, and capacity-building at all levels. In Istanbul, the cost of retrofitting 5 000 buildings with the possibility of saving 20 000 lives could cost USD 200 million (but the cost of a disaster could be far greater). Should retrofitting be financed, and if so, how?

A basic objective of post-disaster reconstruction is to reduce the vulnerability of a devastated region to catastrophe in case of future natural disasters through better spatial planning, building codes and enforcement measures, and civil preparedness. *Compensation or insurance* often provides for reconstruction of an identical house so as to avoid moral hazard questions, lest people who suffered a loss receive a net benefit. This gives people incentives to rebuild quickly (thus diminishing the demand for public services and support), but in ways that make any readjustment of the area all but impossible. A related problem concerns ownership and tenure. In the Northridge, California earthquake of 1999, a large proportion of damaged properties were apartments. Owner-occupiers have an obvious motive to rebuild, but owners of apartments may simply collect for damage and loss, choosing not to rebuild, but thereby altering the mix of housing types available in a community which could affect its social structure and job mobility.

Another moral hazard issue arises if people who were negligent in respect of building codes receive compensation together with people whose houses were in compliance. Such a precedent would do nothing to encourage better enforcement and higher levels of compliance in the future. This pattern can only be broken if compensation is supported by measures that improve enforcement, including perhaps a scheme for mandatory insurance linked to code compliance. Thus, the management of financial resources to aid and rebuild regions may need to reflect risk reduction, rather than promote *in situ* rebuilding of what was lost in an effort to re-house people quickly. *Relocating people to a new area permanently (e.g.*, Mississippi Valley (US), Izmit Region (Turkey) has been managed on a relatively small scale in democratic societies, but is more common on a large scale in autocratic states. Resistance to change that is "top down", without the participation of local authorities and civil society, can lead government to modify or withdraw its proposals;

alternatively, people may choose to move to another region entirely rather than accept the terms of the readjustment plan.

Disaster-relevant planning involves analysis of the vulnerability of different types of regions to disasters based on land-use patterns, infrastructures, etc. Relocation and rebuilding were compromised in Turkey due to the absence of up-to-date geological maps following budgetary and management policies that weakened the cartographic and scientific capacity to produce reliable information at sufficient scale. The problem of territorial adjustment can be exacerbated by the constitutional protection of property rights, which can lead to litigious delays, compromising the implementation of a strategic plan and expropriation procedures. An appellate or administrative body can play a role by adjudicating claims.

The current pattern of urban and regional development ("sprawl", increased settlement in rural and coastal regions) is aggravating the likelihood of property loss and loss of life from floods, hurricanes and earthquakes. As things stand, governments often put short-term development and investment ahead of disaster risk. Millions of people who now live in areas of the south-eastern United States where violent storms are likely have not lived through such an event, and as a result, may be less able to understand the value of measures to reduce storm damage, and less able to cope on their own when a storm does occur. To reduce the vulnerability of places to the risk of natural (or industrial) disaster, the objectives of spatial planning and the pattern of co-operation among local, regional and national levels of government may need to be changed. (In England, 10 per cent of the population and 12 per cent of the agricultural land are in flood-risk areas, where property worth more than 200 Bn. pounds is at risk. Floods in north-west Europe in 1995 forced the temporary evacuation of 250 000 people.)

A recent study of more than 250 cities with at least one terrorist incident between 1993 and 2000 (for a total of 1 326 deaths and 11 762 injuries) concluded that cities which were media and communications centres, and where diverse groups are often in conflict with each other, and which exist at geographical and social junctures or fault-lines, were more at risk (Savitch, 2001). Such cities would have both a high level of social breakdown and a significant capacity to mobilise radical movements through networks and organisations. A recession, combined with rising deficits for state and local governments and mandatory increased funding for security purposes, may handicap cities in the United States in their efforts to address the social and economic conditions which foster cohesion and integration (Simpson, et al., 2002). The impact of economic, spatial and social trends at territorial level on the exposure of communities to different kinds of risk, including terrorism, should be a priority for further study, together with efforts to develop indicators of social and organisation capacity to cope with stresses and shocks.

As the standard of living increases, and as public reliance on technology grows, the tolerance of social disruption and economic loss associated with natural disasters diminishes. Thus, people come to expect that the consequences of disasters can be minimized, and react more critically when they are not. Rather than accept the likelihood of a disaster and adopt some damage-reducing measures that may save property and lives, people may try to live as if the disaster either will not happen, or will not affect them. Public education and communication on a regular basis are necessary.

Major issues at territorial level include:

- More robust measures to manage urbanisation in coastal and flood-prone zones.
- Assess the potential gap between value of losses and insurance cover.
- Develop insurance and compensation rules to facilitate relocation and safer buildings.
- Incentives and constraints to achieve compliance with building codes and planning restrictions.
- Clarify relations between levels of government.
- Increase capacity to plan for reconstruction within a strategic vision for regional development.
- Raise level of public awareness of risk on a sustained, ongoing basis.

Additional sources

Nemetz, P., N. and K. Dushnisky (1994), "Estimating Potential Capital Losses from Large Earthquakes", *Urban Studies* 31, 99-121.

Savitch, H.V., and G. Ardashev (2001), "Does Terror Have an Urban Future?", *Urban Studies* 38:13, 2515-2533.

Simpson, D., *et al.* (2002), "Chicago's Uncertain Future Since 11 September, 2001", *Urban Affairs Review* 38:1, 128-34.

Williams, G. (2000), "Rebuilding the Entrepreneurial City: The Master Planning Response to the Bombing of Manchester City Centre", *Environment and Planning B: Planning and Design* 27, 485-505.

Key issues

7. Lessons learned from nuclear accidents

by Edward Lazo

Although the many consequences of nuclear accidents differ from those of other technological or natural disasters, there are also many similarities. Much experience has been gained in planning, preparedness and management of nuclear accidents through addressing real situations like Chernobyl, or fictitious situations through drills and exercises. Much of this experience is directly applicable in the preparation for and management of other, large-scale disasters.

Planning phase

In the planning phase, it is important to develop an emergency response structure with clear identification of the roles and responsibilities of all organisations and individuals involved. Because no two accident situations are alike, it will be necessary for the response structure to be flexible enough to respond effectively, yet endowed with sufficient resources and responsibility to appropriately address many diverse needs. Because of the unpredictable nature of accident situations, the processes and mechanisms for decision making and decision co-ordination are more important than the structure itself (NEA, 2001a and 2000a).

Emergency and immediate post emergency phase

One of the key lessons learned in planning for and managing accidents in the nuclear area is that effective and efficient communication at the local, national and international levels is extremely important (NEA, 2001a). This applies to all time phases of an accident (i.e. before and during the accident, following the regaining of control, and during the recovery period) and must be built into the emergency response programme at the planning stage. Communication among national organisations with accident response roles is critical to assure effective decision-making and public protection. In addition, it should be recognised that any situation having public health implications, even if they initially affect only a restricted geographic area, will be of interest to the governments of other countries. Foreign governments will be interested not only in the direct health and safety of their citizens in the affected country, but in broader and potentially longer term issues such as trade in commodities and foods (through radiological, chemical or biological contamination, for example), effects on foreign businesses located in the affected area, tourism to the affected area, or travel (by air, rail or car) to and from the affected area.

These concerns were broadly seen within European governments during the Chernobyl accident, and occur within virtually all international nuclear emergency exercises. For these reasons, it is extremely important that national response mechanisms include communication with foreign governments for the efficient transfer of official information. This conclusion has policy and structural implications.

Beyond simple communications, in many situations, it would also be very useful for governments to achieve some level of co-ordination in their decisions. For example, following the Chernobyl accident many governments, particularly in Europe, imposed trade and travel restrictions (NEA 2002). A *priori* discussion of such decisions might have avoided some of the trade disruptions that were at least partially caused by significant differences in the national trade approaches. International organisations have, since this time, established agreements on radiological criteria for trade in food, and there may be useful examples of mechanisms for the establishment of internationally agreed-upon approaches to similar issues in other situations. To achieve this, co-ordination must be a policy goal that is supported by structural mechanisms.

Recovery phase

Although urgent response to accidents must be mechanical and proceduralised, the involvement of stakeholders in post-accident decision making processes is essential (NEA, 2001b; TRUSTNET, 2002). This is particularly true in situations that have or might have lasting human health or environmental consequences. While the health risks associated with exposure to radiation are quite well studied and understood, and can be scientifically estimated for a given exposure situation, the acceptance of a particular level of risk is a social and political issue. Successfully identifying and implementing protection measures that are acceptable to the affected populations thus involves a governmental commitment to public or stakeholder dialogue. In the case of potentially lasting radiological effects, this commitment must be long term. While this may seem to slow the process of recovery, it will inevitably lead to a more enduring solution. Again, this conclusion carries policy and structural implications.

In some cases, it should be recognised that existing administrative organisations may not be best suited to such lasting and diverse commitments, such that structural changes in governmental organisations and decision-making processes may be necessary to allow the level of stakeholder participation that will lead to accepted solutions. As an example, a 1999 criticality accident at a small fuel fabrication facility in Japan resulted in a few hundred members of the public living within a few hundred meters of the facility to be exposed to extremely low levels of radiation, having no discernible health effects. However, in response to public concerns over the government's emergency response to this accident, Japanese nuclear regulatory and emergency response organisations were significantly remodelled, and a series of 20 or so regional emergency response and emergency training centres were established. These actions carried policy, regulatory and monetary costs.

The involvement of stakeholders will also affect governmental need for flexibility. International harmonisation of approaches to certain aspects of public and environmental protection, such as food and commodity standards for trade, is necessary. However, governments will need to ensure the availability of a certain flexibility to address stakeholder needs in their specific situations. For example, discharges from a French nuclear installation provoked public concerns in the late 1990s over cancer among those living in the area. Over the period of several years, the French government established citizens' groups to address issues of concern, including environmental measurement and exposure modelling approaches (CEPN, 2000). These dialogues, which also included the nuclear installation as one of the stakeholders, effectively addressed the concerns of local inhabitants, helped to re-establish trust in government, and led to the establishment of new radiological release criteria for the facility. This would not have been possible had policies, regulations or procedures been too rigid. In the case of Chernobyl, strict radiological criteria for environmental contamination and population compensation were established, soon after the accident, largely without the participation of the affected inhabitants. The adoption of these criteria, among other things, obliged the Belarusian Government to devote approximately 20% of its annual national budget, for over 15 years, to Chernobyl-related expenditures (NEA, 2002; OECD, 1998). The lack of involvement of the populations living in the contaminated territories in the establishment of a long-term recovery plan has contributed to public distrust of government institutions, to pessimism in the future, and to increasing stress-related illnesses. Only now, more than 15 years after this accident, are stakeholder involvement programmes being established to assist affected populations to better understand their situations, to affect their own exposures through lifestyle modifications, and to begin to regain confidence for the future of their children. Thus, early decisions to identify and address stakeholders' concerns can have significant effects, positive or negative, on long-term response efforts and issues.

Liability sharing

As a result of the realisation that the potential costs of nuclear accidents could be very large, and that nuclear accidents could have trans-boundary effects that could also carry large costs, governments of countries using nuclear power have established, as early as 1960, international conventions for the sharing of compensation burdens. The objective of these Conventions is to ensure the adequate compensation of damage caused to persons and to property by a nuclear accident. The Conventions aim also to establish a uniform international system, and to facilitate an international collaboration between national insurance pools if it is necessary for financial security to be made available to meet the possible compensation claims.

Today, three international conventions (the Paris Convention, the Brussels Supplementary Convention, and the Vienna Convention) establish special liability regimes based on the following principles:

- strict liability of the operator (liability without fault);
- exclusive liability of the operator;
- limitation of this liability in amount and in time;
- obligation of the operator to cover his liability by insurance or other financial security.

The Paris Convention has 15 contracting parties (national governments from Western Europe), and the Vienna Convention has 32 contracting parties (national governments from Eastern Europe, Asia and South America). The Paris and Vienna Conventions establish, respectively, that operators of nuclear installations must hold financial security of no less than 700 million and 300 million Special Drawing Rights (SDRs). In addition, the Brussels Supplementary Convention to the Paris Convention (Signed by 12 of the 15 Paris Convention contracting parities) adds that, should the liability needs exceed the operator's insurance, the accident state would contribute up to 300 million SDRs, beyond which the other contracting parties would contribute up to 500 million SDRs. It should be noted that these values have not yet come into force but will do so as a result of the currently ongoing negotiations to update the Paris and Brussels Supplementary Conventions. The fraction of the communal funding contributed by each contracting party is a function of the country's GDP and of their installed nuclear capacity. The conventions also establish the definition of what damages are subject to compensation under the conventions.

Finally, it should be noted that similar conventions exist for shipping accidents involving both petroleum products (the International Oil Pollution Compensation Funds – IOPC Funds), and hazardous chemicals (International Convention on Liability and Compensation for Damage in Connection with the Carriage of Hazardous and Noxious Substances by Sea). In both these cases, operator liability funds are required by the conventions, above which additional funds provided by the conventions become available.

Cost analyses methodologies

In a more analytical sense, methodologies to estimate the costs of nuclear accidents have been extensively studied. Different approaches to the evaluation of costs have been used, including the "willingness-to-pay" approach and the "human-capital" approach. Different approaches to the identification of consequences, such as the probability-weighted aggregation of scenarios, or the use of single, worst-case scenarios, have been employed in various studies. It has been concluded, in the nuclear industry, that the most appropriate assessment methodology to use will depend upon the intended use of the cost estimate results (NEA, 2000b). For example, to estimate compensation aspects, a probabilistic approach may be the most appropriate. The assessment of potential damages may be best represented by a single-scenario approach, assuming a worst credible accident for example. The scope and objective of such assessments should be clearly established before beginning such an exercise, and the cost assessment methodology should match the objective at hand.

References

CEPN (2000), Nord-Cotentin Radioecology Group: An Innovative Experiment in Pluralist Expertise. CEPN-R-269 (EN), November.

NEA (2000a), *Monitoring and Data Management Strategies for Nuclear Emergencies*, OECD/ NEA, Paris.

NEA (2000b), *Methodologies for Assessing the Economic Consequences of Nuclear Reactor Accidents*, OECD/NEA, Paris.

NEA (2001a), *Experience from International Nuclear Emergency Exercises: The INEX 2 Series*, OECD/NEA, Paris.

NEA (2001b), *Policy Issues in Radiological Protection Decision Making: Summary of the 2nd Villigen (Switzerland) Workshop*, January; OECD/NEA, Paris.

NEA (2002), *Chernobyl: Assessment of Radiological and Health Consequences – 2002 Update of Chernobyl: Ten Years On*, OECD/NEA, Paris.

OECD (1998), *Environmental Performance Reviews: Belarus*, OECD, Paris.

TRUSTNET (2002), A *Report of Trustnet on Risk Governance – Lessons Learned*, Heriard Dubreuil G., G. Bengtsson, P.H. Bourrelier, R. Foster, S. Gadbois, G.N. Kelly, N. Lebessis and J. Lochard, Journal of Risk Research, Vol. 5, No. 1, pp. 83-95.

ISBN 92-64-02018-7
Large-Scale Disasters
Lessons Learned
© OECD 2004

Chapter 3

The 1999 Marmara earthquakes in Turkey

by Veda Akgiray, Gulay Barbarosoglu and Mustafa Erdik, Bogazici University, Istanbul, Turkey

On 17 August, 1999, an earthquake of magnitude Mw = 7.6 hit the Kocaeli and Sakarya districts of northwestern Turkey. It was soon followed by another big earthquake of magnitude Mw = 7.2 in the close by area centred around Duzce. The disasters affected an extensive area covering the cities of Kocaeli, Sakarya, Düzce, Bolu, Yalova, Eskisehir, Bursa, and Istanbul. The combined population of these eight cities constitutes about 23% of the national population. The heaviest damage was recorded in Kocaeli, Sakarya, Bolu, and Yalova, whose combined population was about 6% of the national figure (down to 4% after the disaster). Most importantly, the Kocaeli – Sakarya region is considered as the industrial heartland of Turkey.

The disasters resulted in more than 18 000 recorded deaths and more than 50 000 serious injuries. More than 51 000 buildings were either heavily damaged or totally collapsed. Another 110 000 buildings were moderately or lightly damaged. As much as 600 000 people were left homeless. To portray the relative scale of the disaster, the ratio of damaged buildings was at least 4 times higher than that in the 1995 Kobe Earthquake and 12 times higher than that in the 1994 Northridge Earthquake.

The 1999 catastrophe was the worst and biggest natural disaster in the history of the Turkish Republic. Its social and economic impact was commensurately deep. This report studies the socio-economic impact of the 1999 Izmit and Düzce Earthquakes in the Marmara region of Turkey. The following typology of costs is used in the report:

- Macroeconomic and industrial impact.
- Impact on public finances.
- Impact on financial markets and insurance.
- Impact on energy and telecommunication infrastructures.
- Community restoration.

Macroeconomic and industrial impact

Macroeconomic costs

The share in total GNP of the eight cities is about 34.7%, and they produce more than 46.7% of total national industrial product. The most affected cities of Kocaeli, Sakarya and Yalova contribute about 6.3% to GNP and 13.1% to industrial output. The per capita income in the earthquake region is much higher than the national average and, as such, the region also constitutes a significant portion of total domestic demand for consumption.

The earthquake has caused wide-spread physical damage, especially to houses and infrastructure. Economic activity has come to a stop and small– to medium–sized businesses have been particularly hard hit. It is difficult to estimate the direct and indirect costs of a large earthquake. Several studies based on very similar assumptions have provided different estimates. The overall estimated cost in terms of both income loss and national wealth loss is in the range from USD 9 to USD 13 billion according to the State Planning Organization and from USD 6 to USD 10 billion according to the World Bank. A study by the Turkish Industrialists and Businessmen Association (TUSIAD) estimates a total loss of more than USD 15 billion. When the indirect and long-term effects are properly considered, it would not be unrealistic to estimate the total cost at around USD 20 billion, which makes about 9-10% of the GDP in the year of 2000.

Although the numerical estimates of costs are not the same, all of the mentioned studies seem to agree on the following taxonomy of economic costs:

1. Income loss

   a) Direct income loss

      i. Loss in domestic value-added (USD 2 billion)

      ii. Decrease in exports and tourism revenues (USD 1.9 billion)

      iii. Increase in imports (+) (USD 200 million)

   b) Indirect income loss

      i. Increase in interest expenditures (USD 1.3 billion)

      ii. Decrease in tax revenues (USD 1.2 billion)

      iii. Emergency relief expenditures (USD 750 million)

      iv. Supplementary catastrophe taxes (+) (USD 5.7 billion)

      v. Decrease in domestic demand for consumption (N/A)

2. Wealth loss

   a) Direct wealth loss

      i. Costs of repair and reconstruction of houses (USD 4 billion)

      ii. Damage to commercial, service and industrial buildings (USD 4.5 billion)

      iii. Damages in infrastructure (USD 1.7 billion)

   b) Indirect wealth loss

      i. Price declines in financial markets (N/A)

      ii. Loss of investor confidence (N/A)

      iii. Costs of bad credit and rescheduling (USD 25 million)

      iv. Insurance damage payments (+) (USD 700 million)

(The numbers in parentheses are approximately either the consensus or the average estimates in the mentioned studies.)

When the earthquakes happened in August and November of 1999, the Turkish economy was still under the negative effects of the Asian crisis and the Russian crisis. Compared to the previous year, the GNP declined by 7.6% and 4.9% in the third and fourth quarters, respectively. The economy shrank by about 6.1% during 1999. Consumption spending decreased by 2.1% and capital investments declined by 16% from the 1998 levels. During August and September of 1999, the Production Index dropped by 12.1% and 9%, respectively, resulting in an annual drop of 5% from the 1998 level. This is largely attributed to the slowdown in production at the Izmit Petrochemical Refinery (TUPRAS), which was seriously damaged by the earthquake.

A survey jointly conducted by the Crisis Management Centre of the Prime Minister's office and the State Statistics Institute after the earthquakes revealed that, of the workplaces with 10 or more employees, 63% were damaged in various degrees. Capacity utilization dropped from 87% to 51% and only half of the workplaces predicted to reach their normal production capacities in 18 weeks. However, this would certainly depend on customer demand, which would take much longer time to come.

Real estate and construction have always been one of the locomotive sectors of the Turkish economy. Immediately after the 1999 earthquakes, all construction permits in the region were suspended until August of 2000. As a result, national construction activity declined by 12.5% in 1999 and it could not fully recover even until 2002. This may be attributed largely to the reluctance of consumers to invest in new real estate while the memory of the 1999 earthquakes is still live.

The earthquakes seem to have had little or no impact on inflation. This is partly due to decreased purchasing power as a result of the disasters and also partly due to the strict anti-inflationary policy of the central government during the same period.

The share of the earthquake region in total imports is about 15% and it is about 5% of total exports. Due to the impact of the earthquake, exports have declined by 6% and imports have declined by 11% in 1999. Although this has resulted in a 27% drop in the foreign trade deficit, its final impact on the economy is to be interpreted as adverse. The earthquake has also resulted in declines in both the number of tourism revenues and also number of incoming tourists. Tourism revenues have decreased by 28% and the number of tourists by 21%.

Damage to industry

Smaller enterprises, employing up to 10 people, were the hardest hit by the earthquake, losing most of their working capital, facilities and workers.

While the total capital stock and value added of the small enterprises might be relatively limited, their large number could bring their total loss to significant levels. World Bank estimated that about 6 000 small shops (employing less than 5 persons) were severely damaged by the earthquake. The total number of small enterprises (employing 5-10 persons) damaged by the earthquake was estimated to reach 2 100. Insurance coverage for small and mid-sized enterprises was very limited; they are typically undercapitalized and have limited access to funding. About 20 000 small businesses have terminated their operations leaving behind about 140 000 jobless people. Job losses could be as much as 45% of the pre-earthquake labour force in the region. Despite government credit incentives, debt reschedules and assistance for re-building, job losses were not fully recovered in the two years following the earthquake. Several sources have reported small business loses to be in the order of USD 1 billion. The loss of capacity in small and medium-sized enterprises has additional adverse socio-economic effects due to loss of employment, production and business relations with larger firms.

Compared to small businesses, damage to large enterprises was moderate and mostly covered by insurance. However, human capital losses in the industry have been more serious. In the weeks following the disasters, there were disruptions to labour supply due to deaths, injuries, and demotivation. Although there are no published statistics, the more serious problem is known to be the migration of hundreds of qualified and experienced workers out of the area.

The epicentre area can be considered as the home of Turkey's heavy industry, including petrochemical plants and car manufacturers. The major industries are automobile, petrochemicals, manufacturing and repair of motor (and railway) vehicles, basic metals, production and weaving of synthetic fibres and yarns, paint and lacquer production, tire factories, paper mills, steel pipes, pharmaceutical, sugar, cement, and power plants. Many foreign companies have affiliates in the region, including Goodyear, Pirelli, Ford, Honda, Hyundai, Toyota, Isuzu, Renault, FIAT, Ford, Bridgestone, Pepsi, Castrol, Dow Chemical, Shell, British Petroleum, Mannesmann, Bridgestone, DuPont, Akza Nobel, Phillips, Lafarge and Bayer. Damage to industry was more extensive than those in other earthquakes with similar ground motion levels. The damage encompassed cooling tower collapses, damaged cranes; collapse of steel, reinforced concrete framed and prefabricated structures, damage to jetties, and extensive equipment failures. The extent of damage to industry depended on, distance to fault, site conditions, quality of construction, anchorage conditions of machinery and robustness and redundancy of fire fighting facilities. Losses due to extensive business interruption were substantial as compared to the physical damage. The Kocaeli Earthquake provided scientists with a unique opportunity to investigate the performance of industrial facilities subjected to substantial strong ground shaking under near-fault conditions.

Petrochemical industry

An extensive concentration of state-owned petrochemical complexes is located within 5 km of the fault, including TUPRAS, PETKIM and IGSAS. The heaviest damage occurred at the TUPRAS facility, the largest refinery in the region producing about twelve million tons per year. The refinery was working at about 90 per cent of its design capacity and can be considered a modern and efficient plant. The earthquake caused significant structural damage to the refinery itself and associated tank farm with crude oil and product jetties. The consequent fire in the refinery and tank farm caused extensive additional damage. Fire started in one of the naphtha tanks continued for three days endangering the safety of the whole region. Six tanks of varying sizes in the tank farm of 112 tanks were damaged due to ground shaking and fire. There were damage to cooling towers and the port area. Collapse of a 150 m high heater stack on the boiler and crude oil processing unit caused significant damage and started a second fire. The total damage is estimated to be around USD 350 million, mostly covered by insurance.

The Petkim petrochemical facility had limited damage, which included settlement at the port and the collapse of a cooling tower. No damage to the equipment in this facility is reported. The fresh water for the Tupras and Petkim complexes, as well as for several other industries in the region (*e.g.*, the Seka paper factory), is supplied from Sapance Lake via 30 km long pipelines. Fault rupture and soil failure caused extensive damage to pump stations and pipelines at about 20 locations. The failure of the water supply caused problems in controlling the fire at Tupras. Igsas fertiliser plant experienced extensive damage in the administration building. Ammonia processing and packing units and the port facilities were partially damaged. At Aksa chemical industries located in the Yalova region, there was damage in port facilities and storage tanks. All of these facilities also experienced extensive losses due to business interruption.

Automotive industry

There are numerous car and tire factories in the region. The Hyundai factory experienced significant non-structural damage. The Toyota car factory had fault ruptures in its parking lot. There was no structural damage to the steel-framed building. Non-structural damage included collapsed storage racks, transformers and cars on the assembly line. Some automatic machinery in the production lines of these factories suffered from alignment problems. Ford Otosan car factory, which was under construction during the earthquake, experienced significant terrain subsidence and some structural damage.

Pirelli Tires, Brisa Tire and Kordsa tire steel belt and cord company had extensive damage and business interruption.

Other industries

Other industrial facilities in the region include cement plants, steel mills, paper mills, food processing plants, textile and pharmaceutical factories. TUVASAS (the railcar production company), Adapazari Sugar Factory and Asil Celik Steel Production Company have all undergone extensive structural damage. In TUVASAS, a large maintenance building and several small buildings collapsed due to lack of bracing in steel structures. In the sugar factory, a stack and an elevator pipe failed and fell into the sugar-processing facility, partly damaging the facility with extensive damage to the equipment inside. Examples of specific damage include collapse of two cranes at the Mannesmann Boru pipe factory; roof collapse, transformer damage, and silo collapses at the SEKA paper mill; collapse of a steel frame structure and movement of bioreactor vessels at the Pakmaya food processing plant; storage rack collapse, toxic releases from mixing chemicals, and damaged piping at the Toprak pharmaceutical firm; and collapse of liquid oxygen tank support structures at the Habas medical gas facility. Kudos textile factory in Adapazari, Cak textile factory in Akyazi (Duzce) and Ak-Al textile factory in Yalova had extensive damage due to the collapse of the pre-fabricated reinforced concrete factory buildings. Some tanks in Aksa chemical installation in Yalova experienced damage, which was associated with leakage of chemicals. Food processing plants that have experienced heavy damage include Pepsi Co-Uzay Gida (Izmit) and Merko Gida (Yalova). In the Duzce earthquake, Süperlit pipe factory, Akisik appliances factory, Sarsilmaz firearm factories in Duzce and Anlas Anadolu tire factory in Kaynasli were heavily damaged. There were limited damage to the industry in Bolu (for example, Filiz Macaroni Factory and Kelebek Furniture Factory).

Initial estimates of the total damage to the industry ranged from USD 1.1 to USD 4.5 billion. The State Planning Organization estimates the value-added loss in manufacturing at USD 600 to USD 700 million and the value-added loss from the damage to industry at about USD 700 million. Other sources put this loss figure as much as into the USD 2 billion range. For example, according to Kocaeli Chamber of Industry, 214 enterprises (about 19% of all enterprises in the province) reported significant damage amounting to a total of USD 2.5 billion in capital losses. Many major facilities are known to face extensive business interruptions, however the biggest loss will be the loss of qualified manpower. A total of 15 per cent capital loss has been reported for the major state-owned enterprise located in the region. (mainly in TUPRAS, TUVASAS, IGSAS, PETKIM, SEKA and Asil Çelik). The State Planning Organization estimates that the loss of business in these factories alone may have amounted to USD 630 million. Since 1999, estimates have been revised but exact figures are still not available.

Impact on public finance

Taxes and other budget revenues

Kocaeli, Sakarya, Bolu and Yalova, the cities most affected by the earthquake, pay about 15% of the total tax revenues in the general government budget. The earthquakes seem to have decreased this percentage, especially in Bolu, Sakarya, and Yalova. There are various reasons for this drop. The first reason is the slowdown in economic activity in these cities. The second reason is the migration from these areas to other cities. A third reason is the decline in the tax collection ratio in the disaster area. The following tables from the Treasury Department of the Ministry of Finance summarize these:

Table 1. **% Share in total tax revenues**

	1997	1998	**1999**	2000	2001	2002
Bolu and Duzce	0.29	0.29	**0.22**	0.10	0.15	0.23
Kocaeli	17.78	14.25	**14.74**	14.20	16.35	18.16
Sakarya	0.41	0.38	**0.23**	0.16	0.22	0.30
Yalova	0.15	0.13	**0.11**	0.11	0.12	0.10

Table 2. **Tax collection ratios**

	1997	1998	**1999**	2000	2001	2002
Bolu and Duzce	90.67	90.07	**81.72**	58.90	37.00	47.06
Kocaeli	98.26	98.00	**97.35**	93.75	96.77	97.53
Sakarya	85.08	84.31	**65.66**	54.34	35.34	51.01
Yalova	92.08	89.63	**83.12**	74.35	82.35	81.48
National average	**89.80**	**89.53**	**86.77**	**90.54**	**90.33**	**91.30**

The decline in the tax collection ratios (the ratio of collections to acruals), especially in Bolu, Duzce, Sakarya and Yalova, is due to the collection difficulties caused by the earthquake and also due to a government decision to postpone 1999 and 2000 taxes to later years. Most of the numbers seem to be regaining their pre-disaster values by 2002.

To finance the economic burden on the government budget placed by the disaster, a number of tax increases and similar revenue enhancing measures have been taken. Almost the whole spectrum of taxes (personal and corporate income taxes, taxes on interest income, property taxes, vehicle taxes etc.) have been increased and the VAT rate has also been increased from 15% to 17%. These new taxes have been popularly called as "earthquake taxes". (It is interesting to note that, at the time, all of these increases have been declared to be temporary but most of them are still in effect today in 2003.) As a result, the additional taxes have generated a source of about USD 5.7 billion in 2000, which was about 11.7% of total tax revenues in 2000. This percentage has subsequently gone down to 3.3% in 2001. In 1999 and 2000, the government had also taken two major measures to generate new revenues. The first was the so called "paid military service" where a qualified candidate paid 15 000 DM to serve for only 28 days instead of a full term of 18 months. This application generated a revenue of about USD 540 million. The second was a cut in the budget shares of certain institutions such as the Istanbul Stock Exchange, Capital Markets Board and the Radio Television Board and reallocate for earthquake spending. This has generated a total of about USD 850 million.

Foreign credit and aid

After the earthquake, various credits and donations have been obtained from various international institutions and countries:

● **European Union:** Under a financial cooperation protocol between Turkey and EU, a total of € 600 million has been committed by the EU as earthquake credit (TERRA). Of this total, € 450 million is for the reconstruction of damaged infrastructure and, to date, € 145 million has been used. The remaining € 150 million is to support the small and mid-sized businesses and € 35.5 million has been paid so far.

● **World Bank:** In 1999, a credit agreement has been signed for a total of USD 757.5 million and a portion equalling to USD 457.4 million has been used so far.

● **International Monetary Fund:** Under an earthquake credit agreement, a sum of 361.5 million SDR has been obtained in 1999 and it has been fully used.

● **European Council Development Bank:** Between 2000 and 2002, various credit agreements have been signed for a total of USD 234 million, all of which has been used.

● **European Investment Bank:** In 2000 and 2001, credit agreements for a total of € 450 million and the € 428 million portion have been used to date.

From other institutions and governments (Islamic Development Bank, Japanese International Cooperation Bank, and the governments of Spain, Japan, Kuwait and Belgium), an additional amount of about USD 140 million has also been obtained, either as loans or as donations.

Spending

The process of government budgeting was expectedly caught unprepared for a catastrophe of such a big scale as the 1999 Marmara earthquakes. The problem was not only the lack of sufficient reserve funds for emergency spending but also the lack of any well-defined policy of budget revision in emergencies. Nevertheless, the Turkish government was as fast as any unprepared management could be in responding to the financial implications of a large scale disaster. It is hoped that there were important lessons learned from this experience.

According to the Ministry of Finance, total spending for the Marmara earthquakes was about TL 3.900 trillion (about USD 6.4 billion) between August of 1999 and December of 2002. About USD 4.2 billion came from the consolidated budget, USD 615 million from non-budgetary sources and various state enterprises, USD 280 million from domestic and foreign aid and donations, and USD 1.4 billion from foreign credit. The final payback cost of this spending on the government budget is projected to be around USD 14 billion, which is a heavy and unexpected burden on the economy for years to come.

Impact on financial markets and insurance

Bond and stock markets

The bond market in Turkey is primarily a market for Treasury bills and notes, and there is virtually no active corporate bond trading. Despite the very high real rates of interest on Treasury issues for several years, the immediate response of the market in August of 1999 was a sharp increase in interest rates. The increase was 12.3 percentage points in the primary Treasury auction issues and 13.6 percentage points in secondary market trading. The average maturity was also shortened by 34 days. Both the interest rates and maturities returned to their pre-disaster levels within a few months after the disaster.

The Istanbul Stock Exchange, which had a market capitalization of USD 114 billion in July of 1999, was closed for a week after 17 August. During the first three days following the re-start of trading, stock prices dropped by 13.6%, which implies a wealth loss of more than USD 14 billion. This drop in prices should be attributed more to deteriorated investor psychology than any rational response to the economic implications of the earthquake. The following year of 2000 was also a bad year for the stock market and the market closed the year with a market cap of less than USD 70 billion. Consequently, whether the Istanbul Stock Market ever recovered from the impact of the 1999 earthquakes will always remain an unanswered question.

Banking

According to the Central Bank records, it is estimated that, in 1999, the total credit risk of the banking sector (including both private and public banks) in the earthquake region excluding Istanbul was about USD 1.5 billion. The breakdown was about 60% private bank credits and 40% public bank credits. The earthquake naturally gave rise to increased probabilities of default and necessity to reschedule the payments. Considering the fact that about one-third of the credit was extended to businesses and individuals, who were directly affected by the disaster, default risk might have reached serious levels. However, there seems to be no evidence of wide spread defaults in the region and also no significant amounts rescheduled. As of August 2000, exactly one year after the earthquake, the amount of rescheduled debt was only USD 26 million.

Insurance

Despite the huge economic impact of the disaster, the 1999 earthquakes did not seriously affect the Turkish insurance sector. This was largely due to the fact that most of the property in the region was not insured against earthquakes. Furthermore, more than 95% of claims were paid by international reinsurance companies, leaving the domestic insurers with a small burden of only USD 26 million. According to Munich-Re, the total insured losses in the region were in the range of USD 600 to USD 700 million, which is from a total insured value of about USD 7 billion in the region. Expectedly, earthquake insurance premium in the affected region have been significantly increased after the earthquake.

One of the most important lessons learned from the earthquake was probably the heavy burden on public finance placed by the lack of proper disaster insurance protection. Within a few months after the earthquake, this was realized and accepted. Through a World Bank project, a national mandatory earthquake insurance plan has been initiated. On 17 December 1999, a government decree was announced to start a national mandatory earthquake insurance scheme for housing units and the legal governmental obligation for the post-earthquake housing was cancelled. A government-controlled insurance pool called the "Turkish Catastrophic Insurance Pool (TCIP)" has been put in place to transfer the national risk into worldwide risk-sharing markets, managed by international reinsurance companies and backed by substantial capital resources. The government hopes that, in the future, the TCIP can contribute to the quality control of construction through differentiation of premium on the basis of earthquake vulnerability. Several opponents to the plan express that it would be expensive, if not difficult, to find adequate re-insurance capacity, the government would have been much better to insure itself while retaining the existing scheme of post-earthquake housing assistance. Several studies have been started to solve the securitization problem of reinsurance risks to make disaster insurance a viable investment.

It is too early to claim that TCIP has been successful. It has some shortcomings and the moral hazard problem of disaster insurance is yet to be overcome. As a case in point, the premium is far short of representing the actual risks. For example, the premium rate is about 0.2% per annum and 2%

deductible for a flat in Istanbul. These rates should be compared with 0.5% premium rate and 10-15% deductible in San Francisco, which has similar natural risk levels. In Istanbul, for a cap of about USD 17 500, the premium is about USD 35. For comparison, in San Francisco, the average premium would be between USD 600-USD 800. Currently, according to Milli-Re, the penetration of compulsory earthquake insurance in Istanbul is about 26% and the pool has a capacity of about USD 1 billion (including the USD 0.85 billion reinsurance) for settlement of the claims in case of an earthquake. There seems to be a stagnation in the market since the number of new policies is almost equal to the non-renewals of existing policies. In case of a large earthquake, the TCIP today with a pool of USD 1 billion can hardly cover expected claims resulting from about USD 3 billion estimated insured losses (about USD 1.5 billion based on a cap of USD 17 500 per housing unit.) More realistic and efficient pricing mechanisms are needed.

Impact on energy and telecommunication infrastructures

Damage to telecommunication systems

Telephone communication was temporarily lost due to damage to the main fibre optic cable at the fault crossing to the east of Izmit and damage to equipment and batteries in the central telecommunication facilities. The usual telephone congestion occurred due to overloaded lines with extensive private calls. Typical damage to the regional telecommunications systems experienced in previous earthquakes was also experienced in Kocaeli and Duzce earthquakes. Many batteries toppled from racks, were broken and needed to be replaced. A number of trunk connections, local loops and cross bars were damaged due to ground shaking and/or falling structures.

Damage to energy infrastructure

Electricity

The electricity in Turkey is produced by TEAS (in cooperation with independent power plants and industrial power plants) and distributed to main transformation stations, also operated by TEAS. The distribution from these main transformation stations to the cities and industrial facilities is done by TEDAS. Both are state-owned companies. Medium voltage (MV) and low voltage (LV) electric power distribution facilities are owned by TEDAS and affiliated distribution companies.

No damage was reported in power plants owned by TEAS and other independent industrial power plants in the Kocaeli and Duzce earthquakes. The power transmission facilities affected by the earthquake are located in eight provinces (Sakarya, Kocaeli, Bolu, Yalova, Bursa, Eskisehir, Bilecik and Istanbul).

The damages, which occurred in High voltage (HV) transmission substations (Adapazari No. 2, Izmit No. 1 and Kentsa), include breakage of transformer bushings, breakage of surge arresters, damage of disconnections, movement of transformers and damages to substation buildings. This damage caused power blackout in north-western Turkey within minutes of the earthquake. The electricity was generally restored to most areas within three to four days.

On the basis of information provided by TEDAS, about 14 (7% of total inventory) of MV/MV type and 800 (7% of total inventory) of MV/LV type distribution transformers in the affected urban areas have received heavy damage. About 850 km (35% of the total) MV type and about 1 300 km (20% of the total) LV type underground distribution cables in the affected region were damaged. Damage to overhead lines was much less. However, 1 050 (7% of the total) MV type towers and 3 000 (5% of the total) LV type towers had to be replaced.

Natural gas and oil pipelines

The state-owned BOTAS (Petroleum Pipeline Corporation), which covers all oil and gas imports and major distribution pipelines, reported no damage on any of their installations. The Russia-Turkey natural gas pipeline in the region crosses the Izmit Bay at about 30 km west of Izmit and the natural gas pipeline connections to industry and power plants in the affected area was fully operational after the Kocaeli earthquake.

In the Izmit municipal gas distribution system (IGSAS), the only urban gas system in the affected region outside of Istanbul, no damage was reported to the main distribution network. IGSAS reports that about 15% of service boxes (out of a total of 21 000) were damaged due to collapsing nearby houses. Some limited damage to service boxes in Istanbul (especially in Avcilar District) was also reported.

Community restoration

Kocaeli and Duzce earthquakes affected a very large area in the nation's industrial heartland, causing extensive building damage, casualties and displaced people. Most of the victims were urban and upper middle class. Damage to industry and small business were widespread. With about half a million displaced people and total losses reaching USD 20 billion, rehabilitation was not an easy and speedy process.

There was a general outpouring of criticism by public to the existing disaster management system. Almost everybody and every organisation were caught unprepared for relief and help after the earthquake. Due to the unusual effort by both government agencies and voluntary civilian organisations, the results of the relief work were not as bad as might have been feared. However, management chaos, duplication of efforts, and lack of address for responsibility were all observed. As a coordination body in future emergencies, General Directorate of Emergency Management under the prime minister's office was formed in 2000. At the local level, municipalities have formed their own emergency relief and aid teams, and civil defence units for SAR operations have also been organized. Some communities have put together voluntary communication networks. However, a nationwide structure encompassing all distributed organisations is yet to be defined and implemented.

About 120 000 families in need of emergency housing after the earthquake were sheltered, in about equal proportions: in tent cities; in individual tents and public buildings and; friends/relatives and rented houses. Within several days to few weeks, a total of 165 000 tents were distributed to affected people/ families. A total of 162 tent cities containing about 28 000 tents were built. Although initially planned to be temporary, as of August 2000, about 30 000 people were still living in 33 tent cities. Several months after the earthquake about 40 000 prefabricated housing units were erected. In addition, about 130 000 families have received USD 300 million "rent assistance", and about 92 000 homeowners received USD 100 million "Light Damage" repair assistance, according to the Turkish General Directorate of Emergency Management).

An important dimension of the rehabilitation and recovery efforts of Turkey is the Marmara Earthquake Emergency Reconstruction (MEER) project. The MEER project is part of the comprehensive Framework Program that has been prepared by the World Bank in cooperation with UNDP, the European Union, other co-financiers and other donors at a total cost of USD 737.11 million. In summary, the project aims at:

- Creation Of Emergency Management Agencies at national and municipal Levels.
- Creation of efficient disaster insurance schemes.
- Modifications in the Disaster Law, Municipalities Law and Public Tender Law.
- Strengthening of the municipal capability for disaster-resistant development.
- Developing risk-based municipal master plans.
- Establishment of a Land Information System.
- Trauma programs for adults.
- Construction of permanent housing.
- Business rehabilitation.

- Repair of existing housing stock and healthcare facilities.
- Rebuilding and repair of infrastructure and lifelines.

The current stage of development of the project is summarized in the following paragraphs.

Repair and strengthening

For repair/strengthening of medium damaged housing units, the government has extended about USD 5 000 low interest credit to about 50 000 homeowners with long payment terms. Extensive applications of repair and strengthening have been and are being undertaken with varying quality and control.

Permanent housing

Permanent housing units to about 38 000 eligible homeowners have been delivered in 2001. Permanent housing units have 100 m^2 floor areas, located in 2-3 Story R/C apartment blocks. The cost per unit is about USD 20 000. The permanent housing blocks are sited on stiff soil or rock, generally located about 10-20 km outside the urban centres. In addition to government provided housing about 19 000 homeowners and 6 000 small businesses have received about USD 10 000 assistance for construction or purchase of new units. All of this assistance will be paid back with very low interest rates over long terms.

Infrastructure/lifelines

About USD 800 million has been spent for the rehabilitation of urban infrastructure. Regarding the highway system, about 85 bridges in the earthquake-affected region were repaired and strengthened.

Small business

The earthquake hit hardest the small businesses. About 20 000 small businesses terminated, leaving about 140 000 jobless people. These losses are being recovered with government credit incentives, debt rescheduling, social security premium deferrals and assistance for re-building. This is expected to be a long-term undertaking.

Legislation for building design and construction control

Legislation has been enacted to enforce mandatory design checking and construction inspection of all buildings by government-licensed private "supervision firms". Supervision firms are required to hire "expert" professionals and have professional liability insurance. (However, the government has recently waived the requirement for insurance due to problems in getting liability insurance with uncertain coverage of earthquake damage.)

Revision of law on engineering and architecture

A professional qualification "expert" system under certification by chambers of engineers and architects has been established. To start the system, all engineers and architects with 12-year professional experience are awarded with the "expert" title. Current activities are underway to provide professional training to those already awarded with the "expert" title.

New institutions

General Directorate of Emergency Management has been formed directly under the Prime Minister's office. Formal civil defence units for SAR operations have been formed in all provinces and several sub-provinces.

Community training

It was seen that people were not prepared for any type of disaster, either mentally or physically. The result was seen as continual fear and loss of hope for the future. Several universities and NGOs with the support of local agencies have launched training programs to increase community awareness and preparedness. Such activities are naturally concentrated in areas with higher earthquake risks.

Conclusion

The 1999 earthquakes taught everybody that earthquake, or any large-scale disaster for that matter, is a fact of life and people have to design ways to cope with it. Although the classical argument has often cited the lack of financial resources as a reason for being unprepared for disasters and the very high economic cost of the disaster (around USD 20 billion), the events following the 1999 disaster and subsequent research on the issue seem to have shown that lack of awareness and appropriate organisation are the real reasons for all human and economic costs of disasters.

ISBN 92-64-02018-7
Large-Scale Disasters
Lessons Learned
© OECD 2004

Bibliography of recent OECD work in related areas

1. Cross-Sectoral Studies

Economic Effects of the 1999 Turkish Earthquakes: An Interim Report
(Economics Department Working Paper No. 247, 2000)

This paper presents a cross-Directorate Report on the economic, budgetary, regulatory and urban policy implications of the earthquakes which struck Turkey in August and November 1999. It traces the factors underlying Turkey's vulnerability to earthquake damage, along a known active fault line, to deficiencies in risk identification procedures and risk-reduction methods, as well as to the absence of risk transfer and financing techniques. It suggests that these deficiencies may stem from the nature of recent Turkish economic development, which has been driven by the need to assimilate a mass migration from the countryside to the cities and has been associated with extremely high and variable inflation.

The Economic Consequences of Terrorism
(Economics Department Working Paper No. 334, 2002)

This paper reviews the prompt policy response which helped limit the immediate economic impact of the 11 September 2001 terrorist attacks, and discusses the possible long-lasting, if diffuse, macroeconomic repercussions. Three channels of influence are explored: shrinking insurance coverage stemming from the perception of greater risk, higher trade costs possibly affecting international trade, and stepped-up security spending partially rolling back the "peace dividend" of the 1990s. It is argued that, in the absence of new large-scale terrorist attacks, and provided terrorism risk is dealt with efficiently, the net long-run macroeconomic impact is probably tangible but limited.

Methodologies for Assessing the Economic Consequences of Nuclear Reactor Accidents (NEA, 2000)

This report was developed to assist in the assessment of the off-site costs that could follow a nuclear reactor accident, focusing on the methodologies that are used and their implications and boundaries. Approaches to the estimation of the costs of countermeasures, of radiation-induced health effects and of indirect and secondary effects were examined. It was shown that the perspective from which the cost analysis was performed (*e.g.* for accident preparedness and management purposes, for compensation purposes, or for power-generation choice purposes) would have a great effect on the numerical outcome. It was also shown that analyses are very sensitive to their initial boundary conditions, including such things as the accident scenario selected, the type of plant studied, and the amount of radioactive material assumed to be released.

Chernobyl: Assessment of Radiological and Health Consequences, 2002 Update of Chernobyl: Ten Years On (NEA, 2002)

This report provides a scientific assessment of the 1986 Chernobyl accident, and of the human and environmental consequences that persist more than 16 years afterwards. In terms of human health consequences, the report draws on numerous national and international studies to show that, the affected populations are suffering from "accident-related" health affects, not associated with radiation exposure but caused by massive social disruption and its associated stress. However, childhood thyroid cancers show significant statistical increase. There have been approximately 2 000 cases so far, with 3 deaths. No other statistical increase in solid-tumor cancer or leukaemia has yet been seen. Environmentally, it is shown that contamination has reached, in many areas, an ecological stability. As such, contamination will persist for much longer periods than originally expected. Finally, significant progress in national and international capabilities to respond in case of nuclear accidents is noted.

Emerging Risks in the 21st Century: An Agenda for Action (OECD, 2003)

This book explores the implications of these developments for economy and society in the 21st century, focussing in particular on the potentially significant increase in the vulnerability of major systems. The provision of health services, transport, energy, food and water supplies, information and telecommunications are all examples of vital systems that can be severely damaged by a single catastrophic event or a chain of events. Such threats may come from a variety of sources, but this publication concentrates on five large risk clusters: natural disasters, technological accidents, infectious diseases, food safety and terrorism. This book examines the underlying forces driving changes in these risk domains, and identifies the challenges facing OECD countries – especially at international level – in assessing, preparing for and responding to conventional and newly emerging hazards of this kind. It also sets out a number of recommendations for governments and the private sector as to how the management of emerging systemic risks might be improved.

Helping Prevent Violent Conflict: The DAC Guidelines (OECD, 1997 and 2001)

2. Insurance and financial consequences

Policy Issues in Insurance: Insurance and Expanding Systemic Risks
(OECD, 2003)

Analyses the consequences of emerging systemic risks and expanding liability for the insurance market, and market mechanisms, possibly backed by government intervention, to cover emerging systemic risks.

Flood Insurance
(OECD, 2002, available on the Internet at *www.oecd.org/dataoecd/51/9/ 18074763.pdf*)

Note on the consequences of the summer 2002 floods in Europe, the issue of insurability of flooding, the available insurance market mechanisms, and the rationale and tools of State intervention in the management of floods.

3. Emergency preparedness and response

1) Chemical accidents

International Directory of Emergency Response Centres for Chemical Accidents – Second Edition (Environment, Health and Safety publications – Series on Chemical Accidents, available on the Internet at *www.oecd.org/ehs*)

The objective of the international directory is to assist emergency planners and responders to prepare for, and to respond to chemical accidents. It does this by listing response centres that are willing to provide help, upon request by government officials or by a centre in another country. The directory is intended to facilitate access to information and assistance provided by response centres located throughout the world.

OECD Guiding Principles for Chemical Accident Prevention, Preparedness and Response – Second Edition (Environment, Health and Safety publications – Series on Chemical Accidents, No. 10; also available on the Internet)

The OECD Guiding Principles is a comprehensive document to help public authorities, industry and communities worldwide prevent and prepare for accidents involving hazardous substances resulting from technological and natural disasters, as well as sabotage. It addresses the following issues:

preventing the occurrence of chemical accidents and near-misses; preparing for accidents through emergency planning, public communication, etc.; responding to accidents and minimising their adverse effects; and following to accidents, regarding clean-up, accident reporting and accident investigation.

2) Radiation

Experience from International Nuclear Emergency Exercises: The INEX 2 Series (NEA, 2001), as well as the finals reports of the Canadian, Finnish, Hungarian and Swiss regional exercises (NEA, 2001, 2000, 2000 and 1998 respectively)

This report presents the collective experience that has been gained through a series of four international nuclear emergency exercises (INEX), organized by the NEA between 1996 and 2001. Each of these exercises was based on a national-level exercise, on which was superimposed an active international component to exercise and assess national and international capabilities to collect and analyze data, and to communicate and co-ordinate decisions. The report documents the key lessons learned in the areas of real-time information exchange, decision making based on limited information and uncertain conditions, public and media communications, and the effective preparation of exercises. This summary report is supported by four reports each specifically addressing the results of the individual exercises.

Inspection of Licensee Activities in Emergency Planning (NEA, 1998 ; available on the Internet at *www.nea.fr/html/nsd/docs/1998/cnra-r98-2.pdf*)

This report presents information on regulatory activities with respect to emergency planning in member countries of the Working Group on Inspection Practices. To establish the basis for these activities, the report first reviews the legal requirements on licensees and second looks into the extent of the regulatory jurisdiction. The focus of the report is on the third section. It reviews the similarities and differences in inspection practices to evaluate compliance with the requirements over which the regulatory body (RB) has jurisdiction.

Status Report on Regulatory Inspection Philosophy, Inspection Organisation and Inspection Practices (NEA, 2001 ; available on the Internet at *www.nea.fr/html/nsd/docs/2001/cnra-r2001-8.pdf*)

This report includes information from both OECD member countries and non-member countries. Chapter 5 provides a list of the requirements for reporting abnormal occurrences within the plant, the response by the inspection programme and follow-up actions taken. Chapter 6 provides information on the role of the regulator during an emergency response situation. Appendices are included which provide organisational structures of both the regulatory body and the national government.

3) Humanitarian assistance

Evaluation and Aid Effectiveness No. 1 – Guidance for Evaluating Humanitarian Assistance in Complex Emergencies (OECD, 2001)

4. Risk management and communication

OECD Guidance Document on Risk Communication for Chemical Risk Management (Environmental Health and Safety Publications – Series on Risk Management, No. 16)

Policy Issues in Radiological Protection Decision Making: Summary of the 2nd Villigen (Switzerland) Workshop, January 2001 (NEA, 2001)

This report summarises the results of the 2nd Villigen Workshop, which addressed the Better Integration of Radiation Protection in Modern Society. Using a series of case studies as a basis for discussions, the workshop analyzed stakeholder involvement aspects of making radiological protection decisions. Cases involving public health and environmental protection issues (*e.g.* population issues in Chernobyl affected areas, cleanup of former uranium mines and mine tailings in Wismut Germany, resettlement of the Marshall Islands) were used to illustrate situations where stakeholder involvement was essential to arriving at agreed-upon solutions. The policy-level lessons learned and implications from the discussion of these cases are summarised.

Public Information, Consultation and Involvement in Radioactive Waste Management (NEA, forthcoming)

This document presents the results of a questionnaire on stakeholder involvement practice and experience in radioactive waste management. A detailed summary of experiences from a number of countries is given. The collected materials show that there is a fund of experience in waste management organisations and regulatory bodies, reflecting many methods and approaches, some more traditional and others more innovative. Important developments or events are taking place at a rapid rate, however, and the information reported in this survey constitutes, for each respondent, only a snapshot from a specific point in time. This document is thus published with the intention to provide both the practitioner and the non-specialist with a valuable baseline of self-generated, detailed information on stakeholder dialogue, consultation and information practices. It can be used to assess the state of the art in the field as well as to provide an historical perspective when assessing future progress.

The Mental Models Approach to Risk Research – An RWM Perspective
(NEA, forthcoming)

Over the last three decades, several empirical studies have reported insights into ways people interpret risks and make inferences and decisions in situations including risks. Efforts have also been made to apply the results of such studies in policy making. The primary motivation of researchers lies in the fact that democratic political systems have not devised mechanisms to adequately support decision makers in handling divergent views on socially acceptable risks, and to identify the role of experts and various stakeholders in controversial decisions. One specific direction of the above research efforts is based on the study of so-called mental models. This approach assumes that, in situations requiring a decision, people generate mental representations of the problems associated with the decision and use these representations to make inferences and decisions. This report provides an overview of the most important current results of both descriptive and prescriptive studies focusing on mental models.

OECD PUBLICATIONS, 2, rue André-Pascal, 75775 PARIS CEDEX 16
PRINTED IN FRANCE
(03 2004 01 1 P) – ISBN 92-64-02018-7 – No. 53357 2004